Padangsidimpuan International Conference on Applied Science, Engineering and Technology (PIC-ASET 2020)

IOP Conference Series: Materials Science and Engineering
Volume 1156

Padangsidimpuan, Indonesia
10 November 2020

Editors:

As'ari Djohar
Sundarapandian Vaidyanathan
Mustafa Bin Mamat
C-H Lien
Mamun Habib
Volodymyr Rusyn
Mumu Komaro
Mujiarto
Jumadil Saputra
Aceng Sambas

ISBN: 978-1-7138-3425-0
ISSN: 1757-8981

Printed from e-media with permission by:

Curran Associates, Inc.
57 Morehouse Lane
Red Hook, NY 12571

CURRAN ASSOCIATES INC. proceedings.com

Some format issues inherent in the e-media version may also appear in this print version.

Institute of Physics
Dirac House, Temple Back
Bristol BS1 6BE UK

Phone: 44 1 17 929 7481
Fax: 44 1 17 920 0979

techtracking@iop.org

Additional copies of this publication are available from:

Curran Associates, Inc.
57 Morehouse Lane
Red Hook, NY 12571 USA
Phone: 845-758-0400
Fax: 845-758-2633
Email: curran@proceedings.com
Web: www.proceedings.com

TABLE OF CONTENTS

ENGINEERING AND TECHNOLOGY

COMPUTER AND MATHEMATICS

PHYSICS, BIOLOGY, AND CHEMISTRY

1st Padangsidimpuan International Conference on Applied Science, Engineering and Technology (PIC-ASET 2020)

Preface

I want to say thank you to all the Keynote Speakers in The 1st Padangsidimpuan International Conference on Applied Science, Enggineering and Technology (PIC-ASET 2020), for your willingness to become the speakers in this conference. And also to all the participants of The 1st Padangsidimpuan International Conference on Applied Science, Enggineering and Technology (PIC-ASET 2020).

The 1st Padangsidimpuan International Conference on Applied Science, Enggineering and Technology (PIC-ASET 2020), provides an excellent international forum for sharing knowledges and results in theory, methodology, and applications of Applied Science, Enggineering and Technology in theoretical and practical aspects. The conference aims to provides a platform to the researchers and practitioners from both academia as well as industry to meet and share cutting-edge development. The 1st Padangsidimpuan International Conference on Applied Science, Enggineering and Technology (PIC-ASET 2020) was carried out online through the Zoom application with capacity of 1000 people as the participants.

The conference was done on 10th November 2020, and organized by Universitas Muhammadiyah Tapanuli Selatan and Universitas Muhammadiyah Tasikmalaya with the co host are Universitas Wijaya Putra Surabaya, IAIN Padangsidimpuan, Universitas Simalungun Pematang Siantar, Universitas Aufa Royhan Kota Padangsidimpuan, Universitas Muhammadiyah Palangkaraya, and Universitas Asahan. The Zoom application is also equipped with a Zoom breakout feature to support the implementation of contribution and streaming sessions on the Youtube application.

The seminar's was held online because Indonesia and other countries are still experiencing the Covid-19 pandemic, so face-to-face seminars cannot do it. Although the implementation of PIC-ASET 2020 is an online conference, the venue for this seminar is still being held at the campus of Universitas Muhammadiyah Tapanuli Selatan, Padangsidimpuan, Indonesia, as a host in Zoom application.

In the implementation of the 1st PIC-ASET 2020, there are still two sessions: plenary and contribution. There were eight presenters from 4 countries in the Plenary Session, namely the Malaysia, Bangladesh, Ukraine and Indonesia. The Plenary Session will be held from 8 am - 4 pm (Western Indonesian Time). Meanwhile, the Contribution Session was divided into 20 parallel rooms using the Zoom breakout feature, which was attended by 133 presenters.

The keynote speakers' presentations are provided, mainly to show the contribution of Computer, Science, Engineering and Technology towards research and knowledge sharing. We have Five keynote speakers coming from the Majelisdiktilitbang PP Muhammadiyah, Adam Jerusalem, Ph.D.; Universiti Tun Hussein Onn Malaysia, Asoc. Prof. Dr. Mimi Nahariah Azwani Binti Mohamed; Independent University, Bangladesh, Prof. Dr. Md. Mamun Habib; Assistant Professor at Department of Radio Engineering and Information Security, Yuriy Fedkovych Chernivtsi National University, Chernivtsi, Ukraine, Volodymyr RUSYN, Ph.D.

The discussion process in the plenary session activity and contribution occurred in two directions because 1 Moderator and 1 Interpreter accompany each Session of the keynote speaker (the Plenary Session) and Presenter (the Contribution Session). So, the implementation of the discussion becomes more active.

The room of the Plenary Session contained up to 800 participants, while the contribution session reached 208 people (including 123 presenters & 85 participants) in the zoom application. Meanwhile, in the contribution session using the breakout application found in the zoom application. Thus, the discussion process, the feasibility of space for Q&A, becomes more optimal, effective, and efficient. All participants in this activity came from various regions in various countries, such as Malaysia, Bangladesh, India, and Indonesia.

In the implementation of PIC-ASET 2020, the technical difficulties that occur are the internet connection, which sometimes hinders the speakers' delivery of material. However, this is not a big obstacle. During the plenary session, a streaming process was also carried out via youtube (https://www.youtube.com/watch?v=kPccBbYJpxU) so that the Speaker at PIC 2020 presented all information can be seen by the wider community, especially the Computer, Science, Engineering and Technology community.

We would like to thank scientific committee, and reviewers, as well as the committee of the Universitas Muhammadiyah Tapanuli Selatan (UMTS) who have participated in the success of this event so that this event can be held as planned. We also conveyed to the Rector of Universitas Muhammadiyah Tapanuli Selatan (UMTS) who had supported this event both in terms of finance and other supporting facilities.

Editors
Prof. Dr. As'ari Djohar
Prof. Dr. Sundarapandian Vaidyanathan
Prof. Dr. Mustafa Bin Mamat
Prof, Dr. C-H Lien
Prof. Dr. Md. Mamun Habib
Assoc. Prof. Volodymyr Rusyn
Assoc. Prof. Mumu Komaro
Dr.Mujiarto
Jumadil Saputra, Ph.D.
Aceng Sambas, Ph.D.

Organizing Committee

- **Advisor** : Dra. Muksana Pasaribu, M.A
- **Person in Charge :** Aisyah Nurmi, M.Pt; Drs. Putoro Dongoran, M.H; Syawaluddin, M.Si
- **General Chair :** Dr. Mujiarto, S.T.,M.T.
- **Conference Chair :** Muhammad Darwis, M.Pd
- **Co-Conference Chair :** Dr. Abubakar, S.Si, M.Si

Member

- **Nurmaini Ginting, M.Si**
- **Rizky Ariaji, M.Pd.**
- **Rafiqah Amanda Lubis, M.P**

Scientific Committee:

- Prof. Sudrajat Supian
- Prof. Lincolin Arsyad, Ph.D.
- Prof. Dr. Muhtadi, M.Si.
- Assoc. Prof. Dr. Fatma Susilawati
- Dr. Ahmed Rhif
- Muhammad Sayuti, Ph.D.
- Dr. Mujiarto, ST.,MT.
- Dr. Janner Simarmata, M.Kom.
- Dr. Sungkono
- Ir. Sardjito, Ph.D.
- Dr. Eng. Agus Setiawan
- Mohammad Adam Jerusalem, Ph.D.,
- Dr. Ade Gafar Abdullah
- Dr. Mumu Komaro
- Dr. Iwa Kuntadi
- Dr. Janner Simarmata
- Mada Sanjaya WS, Ph.D.
- Jumadil Saputra, Ph.D.

Editorial Committee

- Prof. Dr. As'ari Djohar
- Prof. Dr. Sundarapandian Vaidyanathan
- Prof. Dr. Mustafa Bin Mamat
- Prof, Dr. C-H Lien

- Prof. Dr. Md. Mamun Habib
- Assoc. Prof. Volodymyr Rusyn
- Assoc. Prof. Mumu Komaro
- Dr.Mujiarto
- Jumadil Saputra, Ph.D.
- Aceng Sambas, Ph.D.

Keynote Speakers

- Mohammad Adam Jerusalem, Ph.D
- Prof. Dr. Md. Mamun Habib
- Asoc. Prof. Dr. Volodymyr Rusyn

Organizing Institutions

- Polytechnic Scool of Tunisia, La Marsa, Tunisia
- National Kaohsiung University of Science and Technology, Kaohsiung, Taiwan
- XiangTan University, Xiangtan, China
- Universiti Sultan Zainal Abidin (UnisZa)-Malaysia
- Universiti Teknologi Petronas-Malaysia Vel Tech University, India
- Independent University, Bangladesh
- Yuriy Fedkovych Chernivtsi National University, Chernivtsi, Ukraine
- Universiti Tun Hussein on Malaysia (UTHM), Malaysia
- Universitas Pendidikan Indonesia, Indonesia
- Universitas Islam Negeri Sunan Gunung Jati, Indonesia
- Universitas Wijaya Putra Surabaya
- IAIN Padangsidimpuan
- Universitas Simalungun Pematang Siantar
- Universitas Aufa Royhan Kota Padangsidimpuan
- Universitas Muhammadiyah Palangkaraya
- Universitas Asahan
- Universitas Muhammadiyah Tasikmalaya

Peer review declaration

All papers published in this volume of *IOP Conference Series: Materials Science and Engineering* have been peer reviewed through processes administered by the Editors. Reviews were conducted by expert referees to the professional and scientific standards expected of a proceedings journal published by IOP Publishing.

- **Type of peer review:** The type of peer review carried out in the article selection process at PIC-ASET 2020 uses a single-blind peer-review. Manuscripts will be sent to at least one anonymous reference for contribution, originality, relevance, and presentation. Besides, Reviewers also do not know whose article they are reviewing.

- **Conference submission management system:** All article management processes in PIC-ASET 2020 starting from submission to acceptance / rejection are carried out through the online system on the page: https://pic.um-tapsel.ac.id and submission system by application on https://confbeam.org/2020/pic/kfz/

- **Number of submissions received:** The number of articles submissions in PIC-ASET 2020 Management System is 123 articles.

- **Number of submissions sent for review:** The number of submissions sent for review is 123 articles

- **Number of submissions accepted:** The number of submissions accepted is 28 articles

- **Acceptance Rate (Number of Submissions Accepted / Number of Submissions Received X 100):** Acceptance Rate = (28/123) x 100 = 22,76%

- **Average number of reviews per paper:** The average number of reviews per paper is at least 1 review.

- **Total number of reviewers involved:** Total number of reviewers involved is 20 Reviewers.

Any additional info on review process: An editor first reviews the submitted manuscript. It will be evaluated, whether it is suitable for the 1st PIC-ASET 2020 focus and scope or has a major methodological flaw and similarity score by using Ithenticate software. Furthermore, the manuscript will be sent to at least one anonymous referee for contribution, originality, relevance, and presentation (double-blind peer review). Reviewers' comments are then sent to the corresponding author for necessary actions and responses. Finally, the suggested decision will be evaluated in an editorial board meeting. Afterward, the Editor will send the final decision to the corresponding author.

Utilizing feedback from the peer review process, the Editor will make a final decision. The review process will take approximately 1 to 2 weeks. Decisions categories include:

1. Reject - Rejected manuscripts will not be published, and authors will not have the opportunity to resubmit a revised version of the paper to the 1st PIC-ASET 2020.
2. Resubmit for Review - The submission needs to be re-worked, but with significant changes, it may be accepted. However, it will require a second round of review.
3. Accept with Revisions - Manuscripts receiving an accept-pending-revisions decision would be presented in the 1st PIC-ASET 2020 under the condition that

minor/major modifications are made. Revisions will be reviewed by an editor to ensure necessary updates are made before publication.

4. Accept - Accepted manuscripts will be published in the current form, with no further modifications required. But before that, it must still be presented in the 1st PIC-ASET 2020.

The reviewers' and editors' suggestions would be conveyed to the author, who will then have an opportunity to revise the paper. A manuscript returned to an author for revision can be held for a maximum of two weeks. Finally, after accepted, all articles that are declared accepted by the Editor, must be presented at the 1st PIC-ASET 2020. The publication will then be proposed to the Journal of Physics: Conference Series published by IOP Publishing.

Contact person for queries:

1. Dr. Mujiarto (mujiarto@umtas.ac.id)

2. Budi Hendrawan,M.Pd. (hendrawan_budy@umtas.ac.id)

Individual Selection Soybean (Glycine max L.) The Genotype of Cross Argomulyo and Tanggamus Based on Character of Yield Component

Arvita Netti Sihaloho, Wahyunita Sitinjak and Tioner Purba

Agriculture Faculty, Simalungun University.

* netti.haloho@gmail.com

Abstract. Selection is one of plant breeding techniques to obtain a desired character. The objective of this research was to determine information about the trait variance in F3 lines in terms of yield and yield components, knowing character could be improved through selection to obtain high yield lines.This study was conducted on june until august 2020 in the village Karang Bangun Rambung Merah Kecamatan Pematangsiantar, kabupaten Simalungun 500-550 with an elevation place, use augmented design. 20 x 30 cm plant distance, data observed for each individual plant, and done to find genetic variance, environment variance, phenotif variance, GCV, broad-sense heritability and Coefficient correlation. The result show that the value of heritability all character was moderate until high except the days of harvest was low. The value of GCV was narrow to the days to flowering and harvest, while other character value of the GCV was moderate until broad. Coefficient correlation indicated that Seed weight per plant was positively and significantly correlated with all studied traits except the days to flowering was negative and non-significant. Based on high heritability and broad GCV, selection process on seed weight per plant so 10 selected lines with highest seed weight..

1. Introduction

Soybean crop is nuts highly favored by the people of Indonesia as the ingredients for the food, tempe, tahu. and soy sauce. Soybeans are in almost all the territory of Indonesia, but had not yet been optimal soybean production. One way to increase production by the use of soybean superior varieties.Superior varieties obtained by doing a cross between parent who having different properties so that it can produce the desired superior character[1].

Superior varieties can be obtained through the activities of breeding by doing in the selection on germ plasm which has been available or by segregation populationselection.The development of varieties soybean directed for the repair of productivity and the quality and adaptation to the special environment[2].

Selection is one of the stages of important in the program for the improvement varieties are the level of success depends on the variation that is inherited genetically, yield character components of soybean become the indicator of the yield and have been used as the selection criteria not direct to develop a high yield varieties[3].

The availability of high diversity on characters repaired and inherited toprogeni's are an indication of success of the program. The genetic variation within a population is one important prerequisite determine the success of improved varieties soybean[4].

The yield component character in soy is important character as an indicator of its results and has often used as the indirect selection in order to develop high yield varieties.The diversity of the components needed to select the best character to be the selection criteria or character of being allowed to be fixed .

Said that yield componentcharacter as the number branches per plant, the number pods per plants and weight 100 seeds can be used in the breeding program to select high yield genotype.The purpose of this research is to get information about the diversity of yield component and yield characters of F3 lines, knowing character selection could be improved through in order to obtain high yield lines[5].

2. Methods

This study was conducted on june until august 2020 in the village Karang Bangun Rambung Merah Kecamatan Pematangsiantar, kabupaten Simalungun 500- 550 with an elevation place, use augmented design. 20 x 30 cm plant distance, F3 lines were 300 plants and the both of parents were 100 plants respectively, by 3 replication. Fertilizing done at the time when planting by 50 kg urea, 100 kg SP36, and 75 kg KCl per acre. Covering the irrigation system, plant maintenance weeding, pest control and disease do optimally according to needs.

Data observed for each individual plant, and done to find genetic variance, environment variance, phenotif variance, Genetic Coefficient Variance (GCV), broad-sense heritability and Coefficient correlation. Criteria of broad heritability according to Mc.Whirter (1979) in [1], under the following condition: $H < 0,20$ = low, $0.20 < H > 0.50$ = medium, $H > 0.50$ = high, whilecriteria of Genetic Coefficient Variance (GCV) according to Knight(1979): 0-10% named narrow, 10-20% named medium and > 20% named broad. The observation is done by measuring height plant, the days of flowering, the number of productive branch, the days of harvest, the number of pods per plant, seed weight per plant and weight 100 seeds.

3. Results and Discussion

Based on the results of its research it can be seen that there are genetic diversity the phenotype and the increasing on characters height plant, the number of pods and seeds weight per plant (Table 1).This shows that the selection of can be achieved because of the population has a broad genetic diversity.The breadth of the value of the differences that produced, good genetic diversity the phenotype and shows that there is a big chance to select the properties of desired.The differences that wider at variety of its genotype and phenotype caused by the seeds are used is the seed F3 was still high levels of segregation[1].

Table 1. The Value of Fenotype and Genetic Variance of F3 Lines By Crossing Agromulyo and Tanggamus

Characters	Fenotype Variance (σ^2 p)	Genetic Variance (σ^2_G)
Height plant (cm)	21.55	45.14
The days of flowering (day)	2.24	1.36
The number of branchs (branch)	2.98	0.65
The number of pods (pod)	292.21	198.16
The day of harvest (day)	2.86	0.09
Seed weight per plant (g)	193.04	132.58
weight 100 seed (g)	3.47	1.24

Genetic diversity wider opportunities selection affect the success the higher.Progressively individual in the population of the nature of the more high frequency genes desired, so the opportunity to get better via genotype more selection and, when narrow, genetic diversity so individuals in population tends to uniform, so selection to repair the trait of being less effective.Genetic diversity be great when far lines related, approaching homosigot, and originated from a cross parents different genetic background[6].

Selection can be done if the selected character havinghigh value of heritability because heritability can guide a trait is affected by genetic factors or environmental factors.High value of heritability shows that genetic factors more of a role in controlling a trait than environment factors. The number of pods and seed weight per plant has high value heritability.

Based on the test, obtained that character of height plants, the days of flowering, the number of pods per plant and seeds weight per plantwas high category (Table 2). This shows that characters is controlled more by genetic factors compared with environment factors. Selection of high heritability characters to start in early generation[7], while the characters the number of branches and weights 100 seeds has a value of heritability was medium.

Table 2. The Value of Broad Heritability (h^2) and Genetic Coefficient Variance F3 Linesby Crossing Agromulyo and Tanggamus

Characters	Heritability (h^2)	Note	Genetic Coefficient Variance (GCV)	Note
Height plant (cm)	2.09	High	24.67	Broad
The days of flowering (day)	0.61	High	4.62	Narrow
The number of branchs (branch)	0.22	Medium	30.54	Broad
The number of pods (pod)	0.68	High	18.65	Medium
The day of harvest (day)	0.03	Low	1.66	Narrow
Seed weight per plant (g)	0.69	High	46.23	Broad
weight 100 seed (g)	0.36	Medium	10.39	Medium

The value of the coefficients of the genetic diversity is presented in table 2 .The value of the CGV as to its character height plant, the number of branches and weights seeds per plant is considered to be broad .This shows the diversity of characters more caused by a genetic factor and matter used is a collection of different germ plasm background genetics[8].

Character of height plant, number of branchs, number of pods, seed weight per plant and weight 100 seed have medium until high of heritability and CGV was medium until broad. That is may be used as selection character. High heritability and broad CGV may be cause response additive gene, while narrow CGV because of not additive gene (dominan or epistasis gene).

Beside the parameter genetic, correlation between kuantitativecharacter with yield has significsnt meaning in selection. The result of the analysis kuantitative character to weight seed per plant can be shown in Table 3. The correlation between weight seed per plant character with character of height plant ($r= 0.614^*$), number of branch ($r= 0.666^*$) and number of pods ($r= 0.857^*$) has positive significant.This show that there is relationship inline which meaning more heighterplant and more how many number of branch and number of pods per plant formed. If the correlation between characters were negative non significant, example character of weight seed per plant with the day of flowering ($r= -0.092^*$). This indicated that height plant not followed by quickly the day of flowering. The plant was high tending to have the day of flowering was older or long live plant.

Selection was effective done if parameter genetic character estimate was high, followed correlation was significant to weight seed per plant character. Based on parameter genetic estimate, correlation was significant between number of pods per plant with weight seed per plant so a few lines selected what have highest weight seed per plant by direct selection method. Direct selection method used because of only weight seed per plant has high heritability and broad CGV.

Table 3. Fenotific Correlation of Kuantitative Character of F3 Lines by Crossing Agromulyo and Tanggamus

Karakter	TT	UB	JC	UP	JPTN	BBTN	B100
TT	1	-0.198	0.616*	0.013	0.674*	0.614*	0.431*
UB		1	-0.127	0.084	-0.125	-0.092	-0.026
JC			1	0.190	0.721	0.666*	0.452*
UP				1	0.154	0.071	0.155
JPTN					1	0.857*	0.466*
BBTN						1	0.490*
B100							1

Note: TT= Height Plant, UB= The Day of Flowering, JC= Number of branchs, UP= The Day of Harvest, JPTN= Number of Pod per Plant, BBTN= Weight Seed per Plant, B100= Weight 100 seed

Selection is a breeding process and base on repairing of all plants to find new superior kultivar and genetic variance was broad. Selection of every plant was important done to can be used to next planting process the higher the percentage that genetic purity. Selection of every plant based on special marker each population by looking traits in earlier generations.

Table 4. Selected Lines Based on Weight Seed per Plant Character

Number of Lines	Weight Seed/plant (g)
4	68.53
97	65.50
44	62.74
73	57.32
89	48.89
14	48.72
93	48.72
87	45.49
79	45.11
76	43.45

Selected lines can show on Table 4. The increasing of weight seed per plant influenced by gentic and environment factor. Because of weight seed per plant character is always used as a selection character.

4. Conclusions

The result of analysis heritability for all character had medium until high only the day of harvest had been low. The narrow CGV only for the days of flowering and harvest character, while the other character the value of CGV was medium until broad. Correlation betweecharacter was positive and significant, only the day of flowering was negative and not significant. Based on high heritability and broad CGV, the selection done to weight seed per plant character, so selected 10 advanced lines with the highest weight seed per plant.

References

[1] M. Barmawi, A. Yushardi, and N. Sa, “AGRONOMI KEDELAI GENERASI F 2 HASIL PERSILANGAN ANTARA YELLOW BEAN DAN TAICHUNG,” 2013.

[2] D. Wirnas et al., “Keragaman Karakter Komponen Hasil dan Hasil pada Genotipe Kedelai Hitam,” J. Agron. Indones., vol. 40, no. 3, pp. 184–189, 2012.

[3] Miftahorrachman, “Korelasi dan analisis koefisien lintas karakter tandan bunga correlation and path coefficient analysis of bunch character on fruit set of salak dwarf coconut,” Bul. Palma, no. 38, pp. 60–66, 2010.

[4] A. Krisnawati et al., "Seleksi populasi F5 kedelai berdasarkan karakter agronomis Selection of soybean F5 population based on agronomic characteristics," PROS SEM NAS MASY BIODIV INDON, vol. 1, no. 3, 2015, doi: 10.13057/psnmbi/m010309.
[5] Z. Iqbal, M. Arshad, M. Ashraf, R. Naeem, M. Faheem Malik, and A. Waheed, "GENETIC DIVERGENCE AND CORRELATION STUDIES OF SOYBEAN [GLYCINE MAX (L.) MERRILL.] GENOTYPES," 2010.
[6] R. T. Hapsari, "Pendugaan Keragaman Genetik dan Korelasi Antara Komponen Hasil Kacang Hijau Berumur Genjah," Bul. Plasma Nutfah, vol. 20, no. 2, p. 51, 2016, doi: 10.21082/blpn.v20n2.2014.p51-58.
[7] L. Hakim, "Komponen Hasil dan Karakter Morfologi Penentu Hasil Kedelai pada Lahan Sawah Tadah Hujan Yield," 2017.
[8] Trustinah and R. Iswanto, "Keragaman bahan genetik galur kacang hijau," Pros. Semin. Has. Penelit. Tanam. Aneka Kacang dan Umbi, no. 0341, pp. 465–472, 2012.

Mapping Climate Classification of Oldeman in Agricultural Resources Management in South Tapanuli District

Imelda Sari Harahap*, Irwan Zulfikri Matondang, Suryanto, Erti Kumala Indah and Irmalia Fitri

Faculty of Agriculture, Universitas Muhammadiyah Tapanuli Selatan, Indonesia.

* imelda.sari@um-tapsel.ac.id

Abstract. The Oldeman classification system is very useful in the classification of agricultural land for food crops in Indonesia using the element of rainfall. The criteria are based on the calculation of wet months (BB) and dry months (BK), respectively, with the limits taking into account the chance of rain, effective rain and water requirements for plants. With the Oldeman climate classification, a cropping pattern system can be determined in an area. Based on the research results obtained 4 types of climate in the South Tapanuli region, namely: D1, D2, E1, and E2. The wettest climate type in the Sipirok region is climate type D1 with 4-5 wet months and <2 dry months. And those that are suitable for planting short-lived rice are 3 (three) harvests or short-lived rice is 2 (two) harvests and secondary crops 1 (sastu) harvests. The driest climate types in the East Angkola region are climate types E1 and E2 with the number of wet months <3 and the number of dry months 6-7. What is suitable for planting short-lived lowland rice is 1 (one) harvest and secondary crops 2 (two) times, the second special harvest falls in the dry season. The topographical condition of an area greatly influences the climatic conditions of the area. With a good climate classification and suitable for rice plants without irrigation assistance, the climate types are D1 and D2, namely in Sipirok District, West Angkola District and a Small Part of Saipar Dolok Hole and Batang Angkola Districts. Meanwhile, the East Angkola sub-district can be planted with rice with the help of irrigation, especially in the dry season.

1. Introduction

Climate is one of the characteristics of land that is very difficult to mitigate, so that climate is an important limiting factor in agricultural planning in Indonesia, while suggests that climate is a natural habit that is driven by several element[1] [2]. Climatic elements include: solar radiation, air temperature, air humidity, clouds and precipitation (rain), evaporation (evaporation), air pressure, and wind. Based on the climate description, it can be identified the type of vegetation that grows in that place, whereas to determine which plants can live in a certain climate, it requires more detailed growth requirements and weather information from several decades with average values and distribution patterns throughout the year, and To estimate plant diversity, daily weather information is needed[3].

Climatic classifications can be mapped by collecting data on climate elements for decades. Along with the occurrence of climate change due to global warming, the possibility of changing climate types is very large, while for decision making in the agricultural sector, information on the climate of an area is needed because from climate and rainfall data it can be determined which plants are suitable for agricultural areas. The suitable climate classification used in Indonesia for agricultural zoning is

Oldeman's classification. This classification is widely used for climate suitability for agriculture because the classification is based on the number of wet and dry months in a row for a minimum of 10 years, this is related to the need for plants to get a lot of water from rain, because the water requirements for plants and their processing are different. different depending on the type [4].

The Indonesian government through the Agricultural Research and Development Agency has created a planting calendar that aims to facilitate farmers in selecting planting periods, types of fertilizers, and types of crops. The making of a planting calendar is obtained from data on rainfall, soil types, water sources, and the vulnerability of an area to disasters, which has been socialized by the government and widely applied by farmers. This condition is of course closely related to Oldeman's classification of agro-climatic zoning which has the same goal, namely the selection of the right type of plant and planting time for agriculture. So it is necessary to study the suitability of Oldeman's classification agro-climate zone theory with its implementation in the field to find out the presentation of the use of climate classification information for farmers.

South Tapanuli Regency is an area located in North Sumatra Province and has an agricultural sector that contributes to the region. The southern part of Tapanuli Regency (Tabagsel) consists of 14 Districts, where some of the livelihoods of the population are farmers (+ 80%) and are one food self-sufficient regency in the province of North Sumatra. However, the classification and mapping of climate information is still very low in the district, so that the farming community only uses old habits / traditions, so it is not uncommon for agricultural products to experience crop failure / fuso due to flooding and cropping patterns that are not suitable for climatic conditions in the region. The existence of climate classification mapping in South Tapanuli Regency can certainly help all parties in making policies related to climate, especially the management of agricultural resources.

So from the description above, the researcher is interested in conducting "Mapping of Oldeman's Climate Classification in Agricultural Resource Management in South Tapanuli Regency" so that agricultural productivity in South Tapanuli Regency can be increased and maintained so that the welfare of the farming community can be increased and the need for food self-sufficiency in the region can be fulfilled.

2. Methods

This research was conducted in the Southern part of Tapanuli (Tabagsel) by observing the rainfall data for the last 10 years covering Padangsidimpuan City, South Tapanuli Regency, Mandailing Natal Regency, North Padang Lawas Regency and Padang Lawas Regency.

The main materials used are rainfall data for the last 10 years which are scattered in the Tabagsel region, topographic maps of the area, books and journals relevant to this research. The tools used are Office Stationery, Computer (Hardware), MS Word and MS Excel.

The data used in this study are secondary data obtained from related agencies. The method of collecting rainfall data is purposive sampling, namely the selection of samples with a specific purpose. From a total of five regencies / cities in the Tabagsel region.

Data processing is divided into several stages, namely calculating the Oldeman classification climate type and processing spatial data. Spatial data processing includes: Interpolation to determine the distribution of climate classification Oldeman and overlay to determine the suitability of the agro-climate zone with the planting calendar.

Processing of spatial data in this study is divided into several stages, namely, the initial stage of collecting data on rainfall post coordinates and entered in the ArcGis software so that the coordinates of the location of the rainfall posts can be displayed on a map and processed in the form of a shapfile. The next stage is calculating the Oldeman climate classification value for each rainfall station, and entering it in the rainfall post attribute data. The analysis technique used is interpolation and overlay. Interpolation is filling in data gaps with certain methods from a data set to produce an area-shaped distribution. One of the advantages of GIS is the data manipulating function, which displays the distribution in the form of an area. Interpolation is used to determine the climate distribution of Oldeman's classification, while overlay is an overlap between maps with the same type of data, area, and coordinate system so that information on two maps can be found in one map. Overlay is used to determine the suitability of Oldeman's classification of agro-climatic zones with the prevailing planting calendar

3. Results and Discussion

From the results of processing rainfall data in South Tapanuli Regency for the last 10 years starting from 2008 to 2017 in the six rainfall stations scattered in the area, the results are in the following table.

Table 1. Rainfall Data in South Tapanuli Regency, 2008 to 2017

No	Rainfall station name	Number of BB (Consecutive wet months)	Number of BK (consecutive dry months)	Climate Type
1	Saipar dolok hole	4	4	D1,D2,E1
2	Simagomago	4	2	D1,D2,E1
3	Padang Balangka	3	7	E1,E2
4	Huta holbung	3	5	D2,E1
5	Hutakoje	4	4	D1,D2
6	Arse	3	6	D2,E1

The results of data processing in the table above are interpolated on a map using the GIS (Geographic Information System) method as shown in the figure.

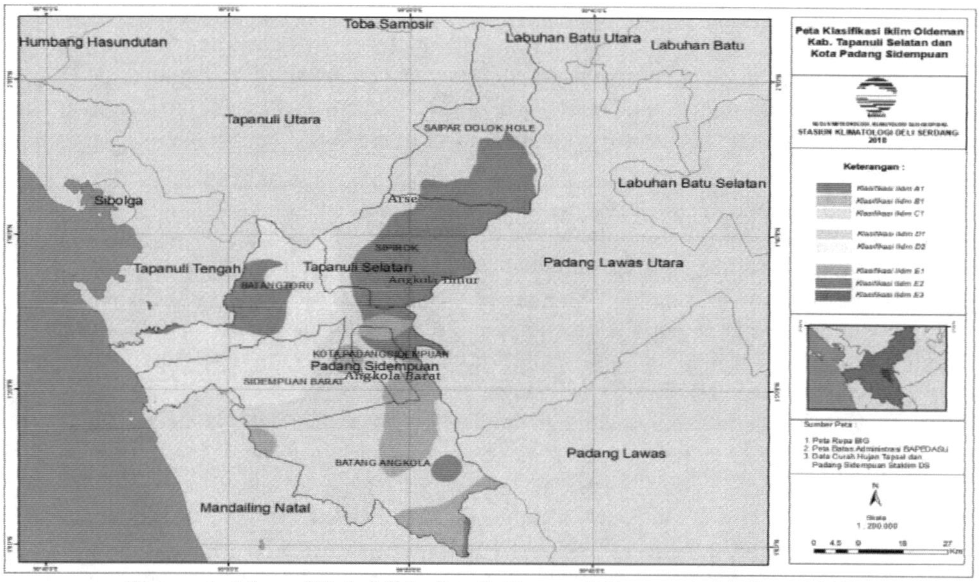

Figure 1. Map of Rainfall in South Tapanuli Regency, 2008 to 2017

The existence of an Oldeman map can help the agricultural sector in knowing adaptation to climate. With the oldeman classification map information, it can be mapped the climate areas suitable for rice in particular. Rice plants (Oryza sativa) can live well in areas that are hot and contain lots of moisture. The average rainfall is 200 mm per month or more than the distribution for 4 months, the desired rainfall per year is around 1500-2000 mm[5].

Based on this information, suitable for rice plants without technical irrigation assistance are climate classification areas C1, D1, and D2. For the climate classification areas E1, E2, E3, rice can be planted with the assistance of irrigation, especially in the dry season[6]. The climate classification areas are D1 and D2 in the six rainfall stations, namely Saipar Dolok Hole District, Sipirok District, West

Angkola District. And Batang Angkola has climate types D2, E1 and Arse has climate types D2, E1. The lowest rainfall is in East Angkola District with climate types E1, E2.

Table 2. Agro-climatic Zones in South Tapanuli Regency, 2008 to 2017

No	Territory	Climate Type	Activity description
1	Saipar dolok hole	D1,D2,E1	Rice has one short life span, high production, very secondary crops, Too dry, maybe only one crop, depending on water availability and irrigation
2	Simagomago	D1,D2,E1	One time short-lived rice, high production, secondary crops, Maybe only one rice or one crop, Too dry, maybe only one crop, depending on water availability and irrigation
3	Padang Balangka	E1,E2	Too dry, maybe only one crop
4	Huta holbung	D2,E1	high production, secondary crops, maybe only once rice or once crops, too dry, maybe only once crops, depending on water availability and irrigation
5	Hutakoje	D1,D2	One time short-lived rice, high production, secondary crops, maybe only once rice or once crops,
6	Arse	D2,E1,E2	Maybe only once rice or once crops, Too dry, maybe only once crops, depending on water availability and irrigation

4. Conclusions

Based on the results of processing and analysis of monthly rainfall data from 6 rain posts in the South Tapanuli Regency, it can be concluded that based on the Oldeman climate classification, there are 4 types of climate in the South Tapanuli region, namely: D1, D2, E1, and E2. The wettest climate type in the Sipirok region is climate type D1 with 4-5 wet months and <2 dry months. And those that are suitable for planting short-lived rice are 3 (three) harvests or short-lived rice is 2 (two) harvests and secondary crops 1 (sastu) harvests. The driest climate types in the East Angkola region are climate types E1 and E2 with the number of wet months <3 and the number of dry months 6-7. What is suitable for planting short-lived lowland rice is 1 (one) harvest and secondary crops 2 (two) times, the second special harvest falls in the dry season. The topographical condition of an area greatly influences the climatic conditions of the area. With a good climate classification and suitable for rice plants without irrigation assistance, the climate types are D1 and D2, namely in Sipirok District, West Angkola District and a Small Part of Saipar Dolok Hole and Batang Angkola Districts. Meanwhile, the East Angkola sub-district can be planted with rice with the help of irrigation, especially in the dry season.

References

[1] As-Syakur, A.R., Iwayan, S., Iwayan, R., dan Inyoman, D. Pemetaan Kesesuaian Iklim Tanaman Pakan Serta Kerentanannya Terhadap Perubahan Iklim Dengan Sistem Informasi Geografi (Sig) Di Provinsi Bali. Jurnal Pastura, Volume 1 No. 1, Agustus 2011, halaman 15-25, 2011.

[2] Kartasapoetra, A.G. Klimatologi: Pengaruh Iklim Terhadap Tanah dan Tanaman. PT. Bumi Aksara. Jakarta, 2004.

[3] Irianto, dkk. Keragaman Iklim Sebagai Peluang Diversifikasi. Pusat Penelitian Tanah dan Agroklimat. Bogor, 2004.

[4] Rafi'I, Suryatna. Meterologi dan Klimatologi: Penerbit Angkasa. Bandung, 1995.

[5] Arronoff, Stanley. Geographic Information System: A Management Perspective. Ottawa : WDL. Publication, 1989.
[6] Sudrajat, A. Pemetaan Klasifikasi Iklim Oldeman Dan Schmist Fergusson Sebagai Upaya Pemanfaatan Sumber Daya Iklim Dalam Pengelolaan Sumberdaya Alam Di Sumatera Utara, 2009.

Acknowledgments
The author expressed his gratitude to the Universitas Muhammadiyah Tapanuli Selatan on basic research through an internal APB grant of research and community service in 2019-2020.

Local Community Perceptions of the KHDTK Mungku Baru Management, Palangka Raya, Central Kalimantan

Ise Afitah[*] and Ardiyansyah Purnama
Muhammadiyah University of Palangka Raya

* ise.afitah@umpalangkaraya.ac.id

Abstract. Mungku Baru Forest Area with Special Purpose (KHDTK) was forest located in the city of Palangka Raya. This KHDTK functions as an area that provides very high economic and ecological value to improve the quality of life of the local community from the economic value of the area as a producing area for fish, resin, rubber latex, and other non-timber forest products. This study aims to determine the local community's perception of the management activities of the KHDTK Mungku Baru. From the research results, it was known that the level of local community perception were included in the low category with an average score of 18.8. Factors on the number of family members, education, and counseling influence the formation of local community perceptions. The low level of local community perception were generally caused by not being involved in planning activities, the level of education that is still inadequate and the level of local community knowledge of the Mungku Baru KHDTK which is still lacking.

1. Introduction

The Mungku Baru Special Purpose Forest Area (KHDTK) functions as an area that provides very high economic and ecological value to improve the quality of life of the community from the economic value of the area as a producing area for fish, resin, rubber latex, and other non-timber forest products. From the ecological value this area is a producer of oxygen as well as a research area for students of the Faculty of Agriculture and Forestry as well as various domestic and foreign parties. With an area of ± 4 910 ha (Four Thousand Nine Hundred and Ten Hectares), KHDTK Mungku Baru has several types of areas such as peat, shellfish, riparian forest, and lowland forest with quite high levels of biodiversity, both vegetation and animals.

In recent years, as the party appointed as manager, the Muhammadiyah University of Palangkaraya has utilized the potential of the KHDTK Mungku Baru area which is a forest area. With the existing potential, UM Palangkaraya developed the area into more than just an ordinary Forest Area. UM Palangkaraya through the Institute for Research and Community Service (LP2M) built and provided quite complete facilities in the KHDTK area. These facilities range from regional entry access, resting places, to information boards.

For the management of KHDTK, UM Palangkaraya also embraces the community around KHDTK to manage it. Co-management which is carried out with a larger portion of management is given to the surrounding community. To find out the need for policies on the management of KHDTK that are right on target, a study of the conditions of the community and the characteristics that shape their behavior is required. Community behavior towards KHDTK can be formed from their perceptions and motivations towards KHDTK, so to find out participatory policies that are right on target, data is needed on the perceptions and attitudes of the people in the area. Public perceptions and attitudes towards KHDTK can be used as a basis for engaging (participation) in the management of KHDTK, so that the conditions for KHDTK can be better in the future and KHDTK can be managed properly. From this research study, of course, it is hoped that it will produce output in the form of information in the form of journals or research reports that are useful for improving the management of KHDTK.

In managing an area of 4 910 ha, the Mungku Baru KHDTK collaborates with the surrounding community. The management methods developed have also had positive impacts on the ecological, economic and social sides. From an ecological perspective, the KHDTK area is preserved, so that its main function can run properly. Apart from that, the animals found in these locations, indirectly, are also well cared for. Likewise for the social and economic side. The community can increase their livelihoods through non-timber forest products obtained from the area.

The problem or question that still needs to be answered is whether the ongoing joint management will have a positive impact in a sustainable or sustainable manner. For this reason, the author wants to examine the factors that influence the sustainability of joint management, namely the perception of the surrounding community in managing the KHDTK, along with the role of the Mungku Baru KHDTK in the lives of the surrounding community.

Based on the formulation of the problem above, the purpose and benefits of this study are to determine the relationship between the level of community perceptions in the management activities of the Mungku Baru KHDTK and to provide information about the level of perceptions of the community around the forest and their role in the management of the Mungku Baru KHDTK so that the Mungku Baru KHDTK manager can use this information. in evaluating and refining KHDTK management policies

2. Methods

2.1 Time dan Place

The location of the research on local community perceptions on the KHDTK management was located in Mungku Baru, Rakumpit District, Palangka Raya City, Central Kalimantan Province. The time for conducting this research was from June 2020 to September 2020.

2.2 Tools and Materials

The materials used in this study were questionnaires and statistical data on the development of KHDTK. The tools used in this research include writing instruments, calculators, cameras, laptops, Microsoft Word software, Microsoft Excel software, and IBM SPSS Statistics 20.

2.3 Metode Pengumpulan Data

2.3.1 Method of Determining the Number of Respondents

Determination of respondents is done by purposive sampling, which is purposive sampling tailored to the purpose of data collection, namely respondents who can or are able to provide data, information and explanations. The target of this research is the people living in the area around the KHDTK, namely the residents of Mungku Baru who are directly adjacent to the KHDTK area. The overall number of respondents interviewed was determined based on the formula of the Slovin method[1], as follows:

$$n = \frac{N}{1 + (N)(e^2)} \tag{1}$$

Information :

n = Number of Samples

N = Total Population

E = Error Tolerance Limit (10%)

2.3.2 Types of Data Collected

The data taken to determine the level of perceptions, of the surrounding community in the management of KHDTK include: a) Primary data obtained by distributing a questionnaire consisting of respondent identity data, including name, age, education, occupation, number of family members , and length of stay. Interview data with the surrounding community and management parties from related agencies. b) Secondary data obtained from agencies related to research and literature study.

2.3.3 Test the Validity and Reliability of the Questionnaire

The validity test is intended to determine the validity of the questions used in this study. The instrument is said to be valid if the correlation value (Spearman Correlation) is positive and the correlation probability value [sig. (2- tailed)] <significant level 0.05. The construct validity was calculated by looking for the correlation of each question with the total score using IBM SPSS Statistics 20 software.

The reliability test is intended to determine the consistency of measuring instruments in measuring the same symptoms. A questionnaire is said to be reliable if the questionnaire can be used repeatedly in the same group and produces the same data. This reliability test uses the Cronbach Alpha coefficient method in the IBM SPSS Statistics 20 software. If the value is positive and the value is close to 1 (has a Cronbach alpha of more than 0.6) it means that the measurement used is reliable.

2.4 Data Processing and Analysis Methods

Data processing and data analysis in this research was divided into several parts, as follows:

2.4.1 KHDTK management activities

KHDTK management activities are described descriptively with pictures after data processing and analysis of the data obtained are carried out, so that an overview of the management system and activities in KHDTK Mungku Baru, Palangka Raya is obtained.

2.4.2 Internal and External Factors

Internal factors and external factors are factors that can influence the community in shaping perceptions and attitudes to decide regarding community participation in KHDTK management activities.

2.4.3 Community Perceptions regarding KHDTK

Public perception regarding the management of KHDTK is measured based on the total score of the questions on the questionnaire using a Likert scale. Measured perceptions of the views of the surrounding community regarding the management of KHDTK on social, economic and environmental life.

3. Results and Discussion

3.1 DecideA general overview of research sites

The Mungku Baru Special Purpose Forest Area (KHDK) is located in Mungku Buru Village and Bukit Sua Village, Rakumpit District, Palangkaraya City, Central Kalimantan Province. Rakumpit District consists of 7 villages, including: Bukit Sua, Gaung Baru, Mungku Baru, Pager, Panjehang, Petuk Barunai, and Petuk Bukit. The area of Mungku Baru Village is ± 29 000 hectares, while the area of KHDTK Mungku Baru is ± 4 910 hectares.

The level of biodiversity in KHDTK, both flora and fauna, is very large, there are about 200 bird species (the largest in Kalimantan), and a variety of other protected animals including orangutans, gibbons, clouded leopards, and others. According to BNF, plants are also forest areas with a higher level of diversity than Sabangau and other BNF managed areas in Purca.

IOP Conf. Series: Materials Science and Engineering 1156 (2021) 012003 doi:10.1088/1757-899X/1156/1/012003

Plants vary. Not only ironwood. There are meranti, tengkawang, restless, benuas, rim, karuing and others. The flora in KHDTK include Ulin (*Eusideroxylon zwageri*), Alau (*Dacrydium beccarii*), Semar Bags (*Nepenthaceae*), Gemor (*Nothaphoebe coriacea*), Ramin (*Gonystylus bancanus*), and Orchid (Orchidaceae).

In the forest there are still many wild animals, such as black bears, mouse deer, wild boar, orangutans, gibbons, proboscis monkeys, bangkui, pangolins, hedgehogs, etc., monitor lizards, crocodiles, especially in the Rakumpit River, there are also many birds. There are 34 mammal species (19 protected), 118 bird species (27 protected), 20 reptile and amphibian species (5 protected) and 28 fish species. There are 5 types of cats including Flat-headed Cat (*Prionailurus planiceps*), Red Cat / Borneo Bay Cat (*Catopuma badia*), Kuwuk Cat / Leopard Cat (*Prionailurus bengalensis*), Clouded Leopard (*Neofelis diardi*), Cat Batu / Marbled Cat (*Pardofelis marmorata*).

3.2 Respondent Characteristics

3.2.1 Age of Respondent

Respondents consisted of various age levels. Respondents were dominated by the age group 37 to 53 years, namely 41.43%. This age is in the range of productive age, statement that the population of productive age ranges from that the population of productive age ranges from 15 to 65 years[2]. The distribution of respondents by age class can be seen in Table.

Table 1. Distribution of respondents based on age

Age (Years)	Quantity (Person)	Percentage (%)
<37	22	36.67
37-53	29	48.33
>53	9	15.00
Total	60	100.00

3.2.2 Number of Family Members

The characteristics of the respondents in terms of the number of family members are dominated by respondents with the number of family members in the vulnerable 4-7 of 56.67%. Family size according to the National Family Planning Coordinating Board [3] is divided into three categories, namely small families (≤4 people), medium families (5-7 people), and large families (≥8 people). The family size referred to by the BKKBN is the size of the nuclear family consisting of a wife, husband and children. The average number of the respondent's family is 4-7 people. The distribution of respondents based on the number of family members can be seen in Table.

Table 2. Distribution of respondents based on number of family members

Number of Family Members	Quantity (Person)	Percentage (%)
<4	19	31.67
4-7	34	56.67
>7	7	11.67
Total	60	100.00

3.2.3 Respondents Education Level

Education has an important role in determining the quality of human resources and the level of welfare so that it can be an overview of the potential resources available in Mungku Baru village. The distribution of respondents based on education can be seen in Table.

Table 3. Distribution of respondents based on education

Education	Quantity (Person)	Persentage (%)
< Primary school	12	20.00
Junior high school to Senior high school	42	70.00
> Diploma to bachelor degree	6	10.00
Total	60	100.00

Respondents generally have a level of education from junior high school to senior high school with a percentage of 70%. This percentage illustrates that in general people have a moderate level of education. Education , especially formal education, is a very valuable asset to get a decent economic life, education is also very influential on the life pattern of each individual, both ways of thinking and behaving[4].

3.2.4 Respondent's Occupation

The types of work in Mungku Baru village are grouped into 4 categories, namely gold miners, fishermen, civil servants and not working. Respondents' occupation is dominated by gold miners by 40%, followed by fishermen's occupation by 38.33%. Comparison of the percentage of respondents by type of work is shown in Table.

Table 4. Distribution of respondents based on counseling

Profession	Quantity (Person)	Persentage (%)
Does not work	8	13.33
Fisherman	23	38.33
Gold Miner	24	40.00
Civil servants	5	8.33
Total	60	100.00

3.2.5 Length of Stay

Respondent characteristics based on length of stay were dominated by respondents with length of stay between 23-46 years, namely 32 people or 53.33%. Complete results can be seen in Table 5.

Table 5. Distribution of respondents based on length of stay

Length of Stay (Yrs)	Quantity (Person)	Persentage (%)
<23	26	43.33
23-46	32	53.33
>46	2	3.33
Total	60	100.00

3.2.6 *Counseling*

There were 21 respondents who received counseling on the Special Purpose Forest Area (KHDTK) of Mungku Baru with a percentage of 35%. The distribution of respondents receiving counseling can be seen in Table.

Table 6. Distribution of respondents based on counseling

Get counseling	Quantity (person)	Persentage (%)
Never	34	56.67
once	21	35.00
Often	5	8.33
Total	60	100.00

3.3 The Local Community Perception on the Management Of the KHDTK Mungku Baru

Perception is an internal process that allows us to select, organize and interpret stimuli from our environment and this process affects our behavior. Perception includes sensing (sensation) through sense organs, organization and interpretation. Broadly speaking, perception is divided into two parts, namely perception of objects (physical environment) and perception of humans. A person's perception is different depending on what the perpetrator is experiencing, both what has happened and is currently happening[5].

The factors that influence perception are functional factors and structural factors. Functional factors are factors that come from needs, past experiences and also other things which are called personal factors. Structural factors are factors that come from outside the individual such as environment, culture, law and values that apply in society and greatly influence individuals in making perceptions [6]. Based on the sum of the scores for the perception questions, the results of the level of perception of the Mungku Baru community towards KHDTK management are shown in Table 7.

Table 7. The level of the local community perception on the management of the KHDTK Mungku Baru

Perception Value	Quantity (Person)	Persentage (%)	Category
29 – 36	13	21.67	High
21 – 28	22	36.67	Medium
12 – 20	25	41.67	Low
18.8			Low

The overall perception of the Mungku Baru community towards the management of KHDTK is in the low category with a score of 18.8. The community around the Mungku Baru KHDTK generally has a poor perception of the Mungku Baru KHDTK and believes that the existence of the Mungku Baru KHDTK does not improve the quality of the environment both in terms of air and climate.

Table 8. Factors that influence the perceptions

Internal & Eksternal Factor	Perception Correlation Coefficient
Age	0.124
Total of Family Member	0.285*
Education	0.241*
Occupation	0.144
Length of Stay	0.112
Counseling	0.793**

* *: Correlation is significant at the 0.05 level (2-tailed)*
** *: Correlation is significant at the 0.01 level (2-tailed)*

From the results of testing the factors that influence perception using the Spearman correlation test at the 95% and 99% confidence levels (Table 8). The factor of the number of families and education has a significant effect on the closeness level of 0.285 and 0.241 on the formation of perceptions.

Meanwhile, the extension factor had a very significant effect, with a level of closeness of 0.793 on the formation of public perceptions.

4. Conclusions

The Mungku Baru community's perception of the management activities of the Mungku Baru KHDTK is in the low category with an average score of 18.8. The factors of the number of family members, education, and counseling influence the formation of local community perceptions. A strategy is needed to increase community perceptions of KHDTK management in order to improve and improve the behavior of the Mungku Baru community so that they are more concerned about the condition of the forest in KHDTK, this strategy can be in the form of increasing counseling on peat, shellfish, riparian forest and lowland forest. In the study of community involvement in KHDTK management activities, it is necessary to consider the needs, aspirations and experiences of the community which are aimed at improving the common welfare so that the benefits obtained from the policy are equally distributed to the community, as well as contribute to improving the social status of the community. Community involvement in KHDTK management activities can be in the form of providing and utilizing part of the KHDTK forest area as a cultivation area managed by the Mungku Baru community. In addition, it is necessary to conduct a study on the behavior of the surrounding community towards KHDTK forest management.

References

[1] Sevilla CG. et. al. Research Methods. Quezon City (PH) : Rex Printing Company, 2007.

[2] Muttaqien A. Paradigma Baru Pemberantasan Kemiskinan: Rekontruksi Arah Pembangunan Menuju Masyarakat yang Berkeadilan, Terbebaskan dan Demokratis. Di Dalam: Raharjo Dawam, penyunting. Menuju Indonesia Sejahtera: Upaya Konkret Pengentasan Kemiskinan. Jakarta: Khanata, Pustaka LP3ES Indonesia. hlm. 3-43, 2006.

[3] [BKKKBN] Badan Koordinasi Keluarga Berencana Nasional. Opini Pembangunan Keluarga Berencana. Jakarta (ID): BKKBN, 2014.

[4] Mursidin. Peran Gender dalam Kehidupan Keluarga Hubungan dengan Tingkat Kesejahteraan Masyarakat. Di dalam: Nasution Zahri dan Hikmah, penyunting. Dinamika peran gender dan disemasi inovasi. Jakarta(ID): Badan Riset Kelautan dan Perikanan, 2009.

[5] Mulyana D. Ilmu Komunikasi: Suatu Pengantar. Bandung (ID): PT. Remaja Rosdakarya, 2005.

[6] Rakhmat J. Psikologi Komunikasi. Bandung (ID): PT. Rermaja Rosdakarya, 2005.

Virulence of *Beauveria bassiana* from Different Carbohydrate Sources Against *Riptortuslinearis* L. (Hemiptera: Alydidae)

Muhammad Agung Permadi*, **Rafiqah Amanda Lubis, Amir Mahmud, Qorry Hilmiyah Harahap, Haryogi Setiawan Sitanggang**

FakultasPertanian, Universitas Muhammadiyah Tapanuli Selatan, Jalan Raja InalSiregar, Padangsidimpuan, North Sumatra 22716, Indonesia

* muhammad.agungp@um-tapsel.ac.id

Abstract. *Riptortuslinearis* is a cosmopolitan pest that is widespread throughout the world. These pests are also polyphagous which meansthis pest has many hosts. The potential of *R. linearis* as a pest needs to be monitored because it is an important pod sucking pest. The entomopathogenic fungi are one of the biological control agents that are often used to control plant pests.One of the entomopathogenic fungi that are often used to control insect pests is *Beauveria bassiana*. Media that contains all the essential nutrients needed fungi for fungal culture in the laboratory. The source of nutrition is a determining factor for the growth and virulence of entomopathogenic fungi because the rate of germination, growth, and sporulation arean indicator of the level of virulence. The research was conducted at the Laboratory of Agrotechnology, UniversitasMuhammadiyah Tapanuli Selatan. This research used a Completely Randomized Design, 5 treatments (5 growth media for *B. bassiana*). The results showed that *B. bassiana* grown in the various media had no significant effect on the mortality of *R. linearis* imago. The lowest LT50 R. linearis was caused by B. bassiana infection grown on rice media.

1. Introduction

Riptortuslinearis (F) (Hemiptera: Alydidae) is a cosmopolitan pest that is widespread throughout the world. These pests are also polyphagous whichmeans this pesthas many hosts. The potential of *R. linearis* as a pest needs to be monitored because it is an important pod sucking pest and spreads across locations and growing seasons, its destructive power is higher than other pod destroying pests, thus indicating a lower economic threshold level. Suction pods can cause seed yield losses both quantitatively and qualitatively (reduce seed quality). The yield loss due to pod sucking pests was up to 80% [1]. The way to damage pod-sucking pests is by sticking the stylet on the surface of the pods and soybeans, causing the soybean seeds to become deflated [2].

Farmer still rely on synthetic insecticides to control pests. Excessive and unwise application of synthetic insecticides will cause unwanted negative effects such as target pest resistance, resurgence, the killing of natural enemies and other useful insects, and environmental pollution. Therefore, an alternative pest control that is efficient, effective, and friendly to the environment is needed by using a biological control agent.

Entomopathogenic fungi are one of the biological control agents that are often used to control plant pests. Entomopathogenic fungi infect by penetrating the cuticle of the host insect, in contrast to bacteria and viruses which must be eaten by the host insect [3]. One of the entomopathogenic fungi that are often used to control insect pests is *Beauveria bassiana*. Several researchers have tested the effectiveness of this fungus against several types of pests including *Spodoptera litura* (Lepidoptera: Noctuidae) [4], mungbean insect pests [5], *Frankliniellaoccidentalis*(Thysanoptera: Thripidae) [6], *Diaphorinacitri* (Hemiptera: Liviidae) [7], *Nezaraviridula* (Hemiptera: Pentatomidae) [8].

Media that contains all the essential nutrients needed fungi for fungal culture in the laboratory. One of the agar media that is often used in the propagation of entomopathogenic fungi in the laboratory is PDA (potato dextrose agar). Several researchers have researched fungal growing media such as cassava starch [9], potato and palmirah tuber [10], and several different carbohydrate sources [11]. The source of nutrition is a determining factor for the growth and virulence of entomopathogenic fungi because the rate of germination, growth, and sporulation is an indicator of the level of virulence.

2. Materials and methods

The research was conducted at the Laboratory of Agrotechnology, UniversitasMuhammadiyah Tapanuli Selatan.This research used a completely randomized design, 5 treatments (5 growth media for *B. bassiana*), and 3 replications. The data obtained were processed with variance and continued with the Duncan test at a 5% significance level.

2.1 R.linearis Rearing

Nymphs and imago were caught from asparagus bean fields at Padangsidimpuan City. The nymph and imago groups are put into a cage. Groups of imago, nymphs, and eggs are reared in the laboratory. The food that is given during the maintenance period is asparagus beans. The asparagus was washed with clean water before placing it in the cage.

2.2 Propagation of Fungi and Preparation of Fungi Suspension

The composition of PDA was potato 400 g, dextrose 15 g, chloramphenicol 1 g, agar 15 g, and 1 l distilled water. Cassava agar media and sweet potato agar media has made the same as the manufacture of PDA.Rice media and corn mediaare made by washing each rice and corn thoroughly, steaming half-cooked, and then cooled in a plastic tray. Media as much as + 50 g is put into clear plastic bags heat-resistant HDPE (high-density polyethylene), sterilized in an autoclave (for 35 minutes with a temperature of 121° C and a pressure of 1 atm). After cooled, each media was inoculated with *B. bassiana*. Inoculation was carried out by spraying 1 ml of fungal conidia suspension with a conidia density of 10^6 conidia/ml into the media bag. All stages were carried out under sterile conditions in laminar airflow. The culture was incubated for 21 days.Each *B. bassiana* from each media was made into a suspension. The conidia density of each suspension was calculated with a Neubauer-improved hemocytometer. This study used a conidia density of 10^8 conidia/ml.

2.3 Inoculation of Entomopathogenic fungi on R.linearis

The application was carried out by spraying the fungal conidia suspension of *B. bassiana* 10^8 conidia/ml on R. linearis imago. Each treatment unit consisted of 11 adult insects sprayed with 2 ml of conidia suspension.

2.4 Observation

The observation parameter was the number of *R. linearis* imago that died after *B. bassiana* application.Observations were made from 1 day after application to 7 days after application.

3. Result and discussion

3.1 Mortality of R. linearis

The research showed that *B. bassiana* grown in various media had no significant effect on mortality of *R. linearis* imago. The research results can be seen in the following table.

Table 1. The mean mortality percentage of *R. linearis* imago after application of *B. bassiana* from various growth media

B. bassiana growth media	Imago mortality (%)
Sweet potato agar media	87.87 A
Rice media	84.84 A
Cassava agar media	81.81 A
Corn media	78.78 A
PDA	78.78 A

Note: Numbers followed by the same letters in the same column show no significant difference based on Duncan (5%).

The Source of nutrients contained in the growth media is a determining factor for the growth and virulence of entomopathogenic fungi. Because of the rate of germination, growth, and sporulation is an indicator of the level of virulence. [12] stated that the type of fungus, the origin of the isolate, genetic differences, the growth medium influenced the entomopathogenicin infecting insects. However, it is different from this study, where the growth media did not affect the fungus *B. bassiana* in infecting *R. linearis* imago. All the growth media tested in this study were able to provide the nutrients needed for the growth of *B. bassiana*, so that there was no significant difference in *R. linearis* mortality caused by *B. bassiana* from different growth media.

The infection mechanism of entomopathogenic fungi begins with the attachment of the fungal conidia to the insect cuticle [3]. Then the spores germinate and penetrate the insect's body. In the next stage, the fungus grows and develops into the insect's blood (hemolymph). The fungus will accelerate reproduction by separating the hypha body to fight insect resistance. At the same time, the antibiotic toxins produced by the fungus both weaken and rapidly kill insects. Furthermore, the hyphae will grow and fill the entire insect body. When the fungus begins to develop, the insect shows symptoms of illness, such as uncoordinated movement and consequently the insect's death.

At the beginning of the death of the test insects, there was no sign of mycelium growing on the test insects' organs. The new fungal mycelium appeared for the first time after the test insects died. Mycelia begin to grow and are found in the articulation organs, especially the legs, then the oral organs then develop in the thoracic part. The articulation organs, including the joints of the legs, thoracic, mouth, and abdominal joints are very flexible areas so that they are easier to penetrate by fungal conidia.

Table 2 LT_{50} imago *R. linearis* after application of *B. bassiana* from various growth media

B. bassiana growth media	LT_{50} (days)
Sweet potato agar media	4.66
Rice media	4.46
Cassava agar media	4.54
Corn media	4.71
PDA	5.00

The mortality of *R. linearis*imago caused by fungal infection takes time because the fungus requires several stages of the process to infect and kill insects, namely conidia attachment to the insect's body, germination, penetration, invasion, and colonization in the hemocoel, tissues, and organs.The time for each of these stages varies depending on the type of fungus, host, and environment. The lowest LT_{50} of imago *R. linearis* was caused by infection with *B. bassiana* grown on rice media. The lower the LT_{50} imago *R. linearis*, the better the results in controlling these insects. This means that *B. bassiana* grown in rice media has a faster rate of infecting and killing *R. linearis* imago compared to *B. bassiana* grown in other test media.

4. Conclusions

The results showed that *B. bassiana* grown in the various media tested had no significant effect on mortality of *R. linearis* imago. The lowest LT_{50} of imago *R. linearis* was caused by infection with *B. bassiana* grown on rice media.

References

[1] M. S. Y. I. Bayu, "Tingkat seranganberbagaihamapolong pada plasma nutfahkedelai," in Pros Sem NasMasyBiodivIndon., Vol. 1, No. 4, pp. 878-883, 2015..

[2] M.M. Rahman, U.T. Lim, "Evaluation of mature soybean pods as a food source for two pod-sucking bugs, *Riptortuspedestris* (Hemiptera: Alydidae) and *Halyomorphahalys* (Hemiptera: Pentatomidae)," in PLoS ONE., vol. 12, no. 4, pp. 1-16, 2017.

[3] D. Rai, V. Updhyay, P. Mehra, M. Rana, A.K. Pandey, "Potential of entomo-pathogenic fungi as biopesticides," in Ind. J. Sci. Res. and Tech., vol. 2, no. 5, pp. 7-13, 2014.

[4] D. N. Erawati, I. Wardati, and S. Humaida, "Potential of *Beauveria bassiana* Lowland Isolates against *Spodoptera litura* in Tobacco Plant," in IOP Conf. Series: Earth and Environmental Science, 2018, pp. 1–7.

[5] M. S. Y. I. Bayu and Y. Prayogo, "Field efficacy of entomopathogenic fungi *Beauveria bassiana* (Balsamo.) for the management of mungbean insect pests," in IOP Conf. Series: Earth and Environmental Science, 2018, pp. 1–9.

[6] Y. Gao et al., "Potential use of the fungus *Beauveria bassiana* against the western flower thrips *Frankliniellaoccidentalis* without reducing the effectiveness of its natural predator *Oriussauteri* (Hemiptera: Anthocoridae)," Biocontrol Sci. Technol., vol. 22, no. 7, p. 803_812, 2012.

[7] M. A. Permadi, R. Anwar, and T. Santoso, "Pemanfaatancendawan*Beauveria bassiana* (Bals.) Vuill. sebagai Miko-Insektisida terhadap kutu loncat jeruk *Diaphorinacitri* Kuw. (Hemiptera: Liviidae)," BioLink J. Biol. Lingkungan, Ind. Kesehat., vol. 4, no. 1, pp. 82–88, 2017.

[8] M. A. Permadi,R. A. Lubis, L. S. Siregar, "Virulensi beberapa isolat cendawan entomopatogen terhadap nimfa kepik hijau *Nezaraviridula* Linn.(Hemiptera: Pentatomidae)," in Jurnal Agrohita, vol. 2 no. 2, pp. 52-60, 2018.

[9] C. K. Kwoseh, M. Asomani-Darko, and K. Adubofour, "Cassava starch-agar blend as alternative gelling agent for mycological culture media," Bots. J. Agric. Appl. Sci., vol. 8, no. 1, pp. 8–15, 2012.

[10] S. Martyniuk and J. Oroń, "Use of Potato Extract Broth for Culturing Root-Nodule Bacteria," Polish J. Microbiol., vol. 60, no. 4, pp. 323–327, 2011.

[11] M. A. Permadi, Mukhlis, B. S. Samosir, D. Y. Siregar, M. Wayni, ". Physiology Characterization of Entomopathogenic Fungi *Beauveria bassiana* and *Metarhiziumanisopliae* on Different Carbohydrate Sources," in J. Phys.: Conf. Ser. 1477 072007.

[12] Y. Prayogo, "Efikasi cendawan entomopatogen *Beauveria bassiana* Bals. Terhadap kepik hijau (*Nezaraviridula* L.)," in Jurnal HPT Tropika, vol. 13, no. 1, pp. 75-86, 2013.

Acknowledgment

Thanks to Universitas Muhammadiyah Tapanuli Selatanfor providing research funding through the APB Internal Grants UM-Tapsel 2019 so that this research can be carried out.

Dormancy Breaking of Kimalaka Seeds (Phyllanthus emblica L.) at Various Concentrations of Sulfuric Acid

Rasmita Adelina[1], Muhammad Nizar Hanafiah Nasution [1*], Darmadi Erwin Harahap[2] and Jihadul Fadhli Nasution[1]

[1]Agrotechnology Study Program, Faculty of Agriculture, Graha Nusantara University Padangsidimpuan, Indonesia.
[2]Agrotechnology Study Program, Muhammadiyah University Tapanuli Selatan, Padangsidimpuan, Indonesia.

* nizarhanafiah.12@gmail.com

Abstract. The kimalaka plant (Phyllanthus emblica L.), known as "balakka", by the people of North Sumatra, with the English translation is called indian gooseberry [1] and is called “popok melaka” in Malaysia [2]. The kimalaka plant is a commodity with great potential to be developed, one of which is due to the high content of vitamin C so that the kimalaka plant has many health benefits. Especially in the Padangsidimpuan area and its surroundings, this kimalaka plant is more widely known as a mixture of traditional cooking spices, with the name of the food being called "holat".The widespread conversion of land into smallholder plantations with commodities such as rubber and oil palm is one of the threats to the extinction of the kimalaka plant. Lack of information about the kimalaka plant is also the cause of the underdevelopment of this kimalaka cultivation. Kimalaka seed with a very hard seed coat is a factor causing kimalaka seed dormancy. So before planting, there must be a treatment to break dormancy. This study aims to see and observe the effect of various sulfuric acid concentrations to break the dormancy of kimalaka seeds.This study used a non-factorial randomized block design (RBD) with one factor, namely the concentration of sulfuric acid with 4 levels, namely: A0 = Control, A1 = 4%, A2 = 6%, A3 = 8%. The observation variables were the percentage of germination (%), plant height (cm), and the number of leaves.Based on the results of this study, it was found that the treatment of sulfuric acid concentration was significantly different in all observation variables. The treatment that gave the best effect on all observation parameters was sulfuric acid at a concentration of 8%. In future research,it is suggested to increase the sulfuric acid concentration.

1. Introduction

Phyllanthus emblica, known in Indonesia as kimalaka [1]. The people of North Sumatra call this plant "balakka". In English, this plant is called Indian glooseberry [2]. while in Malaysia it is called “Popok Melaka” [3] and Thailand is known as Ma-kham-pom [4]. The Indian state called the milan plant by various names, for example, Aonia, , Amla, Amlika, Dhotri, Embelica, and Usuri [5]. Phyllanthus embelica generally grows in tropical and subtropical areas including in India, China, Indonesia, Peninsular Malaysia, Thailand [4]. Kimalaka fruit contains a high source of vitamin C [6]. According to [7], every 100 grams of kimalaka fruit contains approximately 600-1300 vitamin C.

The kimalaka plant is a commodity with great potential to be developed, one of which is due to the high content of vitamin C so that the kimalaka plant has many health benefits. Especially in the Padangsidimpuan area and its surroundings, this kimalaka plant is more widely known as a mixture of traditional cooking spices, with the name of the food being called "holat".

The widespread conversion of land into smallholder plantations with commodities such as rubber and oil palm is one of the threats to the extinction of the kimalaka plant. Lack of information about the kimalaka plant is also the cause of the underdevelopment of this kimalaka cultivation. Kimalaka seed with a very hard seed coat is a factor causing kimalaka seed dormancy. So before planting, there must be a treatment to break dormancy.

The hard seed coat is impermeable to water and air, which prevents the germination process of the seed. To speed up the process of breaking dormancy and the germination process of thick and hard-skinned seeds, several ways must be done, one of which is by soaking the seeds in chemical solutions such as sulfuric acid (H2SO4), nitrat acid (HNO3) and hydrochloric acid (HCl) [8].

Sulfuric acid is a solid mineral (inorganic) acid. This substance dissolves in water at all ratios. H2SO4 can break down the cell wall components of the seeds, so that the cell walls are more permeable and the water absorption process in the seeds runs well [9]. Sulfuric acid (H2SO4) at the appropriate concentration can soften the waxy layer on the hard and thick seed coat [10]. The cell wall is composed of cellulose microfibrils which consist of polysaccharides. Treatment with sulfuric acid can break the bonds of cellulose microfibrils causing the cell wall to be more permeable, so that water and oxygen can easily enter the seed cells. Water and oxygen that enter the seed cells are needed for embryonic respiration in the seeds [11]; [12].

Based on the above background, this study aims to study and see the effect of sulfuric acid immersion at various concentrations on germination and growth in kimalaka seeds. The implementation of this research is one of the efforts to solve the kimalaka plant cultivation problem, especially in the kimalaka seed dormancy problem.

2. Methods

This research was conducted in October-December 2018. The location of the research was carried out in the screen house of the Faculty of Agriculture, Graha Nusantara University, Padangsidimpuan, with an altitude of about 480 m above sea level.

The materials used in this study were Kimalaka seeds (Phyllanthus emblica L), sulfuric acid (H2SO4), topsoil, water, and sand. The tools used in this study were knives, trays, hoes, wood, digital scales, labels, cameras, and other stationery.

This research was conducted in a non-factorial randomized block design (RBD) with different concentrations of sulfuric acid and the same immersion time for 45 minutes.The treatment of this study consisted of 4 levels, namely ;A0: Control, A1 = 4% sulfuric acid, A2 = 6% sulfuric acid, A3 = 8%Sulfuric acid. Observation variable in this study : percentage of germination (%) plant height (cm) , and number of leaves (strands).

3. Results and Discussion

3.1 Germination percentage (%)

Based on the analysis of variance and continued with the 5% LSD test, it can be seen that the application of sulfuric acid with different concentrations has a significant effect on kimalakaseed germination.The average percentage of plant germination can be seen in the table1.

Based on Table 1, from the results of the 5% LSD test, it can be seen that the percentage of plant germination has a significant effect, it can be seen from the A0 treatment (control) significantly different from all treatments A1 (H2SO4 4%), A2 (H2SO4 6%), and A3 (H2SO4 8%). From the results of the research, the best treatment was treatment A3 (8%) with a percentage of germination 61,108%. The results of this study are supported by the statement of [13] which states that strong acids are very effective in breaking dormancy in seeds that have a hard seed coat structure.

Table 1. Average Percentage of Kimalaka seed germination (7 WAP)

Sulfuric acid concentration (%)	Percentage of germination(%)
0	**20.83 a**
4	38.88 b
6	51.38 c
8	**61.10 d**

Note: The numbers followed by the same letter in the same column show no significant difference based on a 5% variance analysis.

3.2 *Plant Height (cm)*

Based on the results of analysis of variance at the age of 5 to 8 weeks after planting (WAP), it was found that the treatment of sulfuric acid with different concentrations on plant height parameters showed a significant effect. The average plant height can be seen in table 2.

Table 2. The Average Kimalaka Plant Height (cm)

Sulfuric acid	Height Plant (cm)			
concentration (%)	5 WAP	6 WAP	7 WAP	8 WAP
0	1.18 a	2.36 a	3.43 a	4.81 a
4	2.21 b	3.02 b	5.50 b	6.36 b
6	2.70 b	3.11 b	6. 09 b	7.00 b
8	3.18 c	3.76 c	6.75 c	7.98 c

Based on the analysis of Table 2, from the results of the LSD 5%, it can be seen that the plant height produces a significant difference. Treatment A0 was significantly different from all treatments, treatment A1 was significantly different from treatment A0, and A3. but not significantly different from treatment A2. Treatment A3 was significantly different from treatment A0 and A1.

Based on the research results, it is known that the best average kimalaka plant height is found in treatment A3, while the lowest treatment is in treatment A0. Plant height is closely related to the rate of germination. If the rate of germination is fast, the seeds will grow quickly so that the plant will grow taller. In addition, the increase in plant height is influenced by the water content around the medium where the seeds are grown. According to [14], water is an important condition for the continuity of plant growth.

Kimalaka plant height increase is indeed influenced by the conditions of the seedlings. The better the treatment given, the better the growth of the kimalaka seedlings, so the better the height of the seedlings. This is in accordance with the research of [15], which shows that the seeds that have germinated will direct their growth for root extension in obtaining more water and nutrients which will be used for growth and development.

3.3 *The Number of Leaves (Strands)*

Based on the analysis of variance, it was found that the treatment of sulfuric acid with different concentrations had a significant effect on the number of leaves of the kimalaka plant.

Observation of the leaf number parameters in Table 3 above can be seen that treatment A0 is significantly different from treatment A1, A2 and A3. The highest average number of leaves was found in treatment A3 (8% concentration), namely 114,666 leaves and the lowest was in treatment A0 (control), namely 48,666. The treatment of sulfuric acid with different concentrations showed that the higher the concentration, the more the number of leaves.

The treatment of sulfuric acid with different concentrations of A3 (8%), namely 114,666 leaves was significantly different from other treatments due to optimal plant metabolism at a concentration of 8% starting from the germination percentage and plant height so that leaf growth increased. This is in accordance with the statement of [16], that the plant metabolic process will increase plant growth which is relatively more perfect in plant growth, which will promote better growth, including an increase in the number of leaves.Supported by the opinion of [14] and [17], in the seeds there is a food reserve which will later be overhauled at the germination metabolic stage, the better the quality of the seeds is followed by a good food reserve, then the germination metabolism will be optimal.

Table 3. The Average Number of Leaves (strands)

The sulfuric acid concentration (%)	The Average number of leaves (strands)
0	48.66 a
4	60.66 b
6	82.83 b
8	**114.66 c**

Note: The numbers followed by the same letter in the same column show no significant difference based on a 5% variance analysis.

4. Conclusions

The sulfuric acid treatment with different concentrations had a significant effect on the germination percentage, plant height, and number of leaves. The treatment with the best effect was obtained at the 8% sulfuric acid treatment and the lowest treatment effect was on the control treatment.

References

[1] Uji, T. Review:KeanekaragamanJenisBuah-Buahan Ash Indonesia dan Potensinya. Puslit Biologi LIPI. Jurnal. Vol: 8(2):157-167, 2006.

[2] Bhandari, P.R., and M.A. Kamdod. Emblica officinalis (Amla) : Areview of Potential therapeutic application . International Journal of Green Farmacy. Vol. 6: 257-269, 2013.

[3] Khan, K.H. and K. Khan. Roles of Emblica officinalis in Medicine - A Review. Botany Research International Vol.2, No 4 pp. 218-228, 2009.

[4] Charoenteeraboon, J., Ngamkitidechakul, C., Soonthornchareonnon, N., Jaijoi, K., Sireeratawong, S . Antioxidant ActiviesofThe Standardized Water Extract from Fruit of Phyllanthusemblica Linn. Songklanakarin Journal of Science and Technology. Vol. 32 (6):599-604, 2010.

[5] Nayaka, D.G. Propagation Studies in Aonla (Phyllanthus emblica L. Thesis. Department of Horticulture College of Agriculture. University of Agricultural Sciences. Dharwad, 2006.

[6] Qureshi, S. A., Asad W., Sultana, V. The Effect of PhyllanthusemblicaLinn on Type- II Diabetes, Triglycerides and Liver- Specific Enzyme. Pakistan Journal of Nutrition. Vol. 8(2):125-128, 2009.

[7] Yulistyarini, T., Ariyanti, E.E., Yulia, N.D. Jenis-jenis Tanaman Buah yang Bermannfaat untuk Usaha Konservasi Lahan Kering. Prosiding Seminar Hari Cinta Puspa dan Satwa Nasional. Kebun Raya Purwodadi-LIPI. Purwodadi. Pasuruan. JawaTimur, 2000.

[8] Purnomosidhi P., J. M. Roshetko, A. Prahmono, A. Suryadi, I. N. Ismawan, and M. Surgana. Perlakuan benih sebelum disemai untuk beberapa jenis tanaman prioritas kehutanan, multiguna, buah-buahan, dan perkebunan. Presowing treatments for some priority timber and multipurpose tree species, fruit species, and estate crops. Lembar Informasi Ag For no. 4 Februari. Bogor, Indonesia: World Agroforestry Centre (ICRAF) Southeast Asia Regional Program, 2013.

[9] Suyatmi, E., D. Hastuti dan S. Darmanti. Pengaruh Lama Perendaman dan Konsentrasi Asam Sulfat (H2SO4) terhadap Perkecambahan Benih Jati (Tectona grandis Linn.). F.MIPA, UNDIP, 2008.

[10] Sagala, J., Perlakuan Benih cendana Dengan Air, asam Sulfat, GA3, Jurnal Departemen Kehutanan, Bogor, 1990.

[11] Wareing, P. F. dan I. D. Phillips. Growth and deffentiation Plants, 3`d edition, Pergamon Press, Chicago, 1989.

[12] Sumanto dan Sriwahyuni. Pengembangan Perlakuan Benih Terhadap Perkecambahan. Pusat Penelitian Dan Perkembangan Tanaman Industri, 1993.
[13] Schmid L. Pedoman Penanganan Benih Tanaman Hutan Tropis dan Sub Tropis (terjemahan) Dr. Mohammad Na'iem dkk. Bandung, 2002.
[14] Sutopo, L. Teknologi Benih. Buku. Rajawali Pres. Jakarta, 2004.
[15] Isbandi. Pertumbuhan dan Perkembangan Tanaman, UGM Press. Yogyakarta, 1989.
[16] Villers, T. A . Seed Dormancy,p 220-282. In T. T . Kozlowski (ed). Seed Biology. Vol II. Academic Press. New York, 2006.
[17] Usmanji, C. E . Studi Dormansi Benih dan Berbagai Cara Pematahannya . Fakultas Pertanian. IPB. Bogor. Hal 63, 1990.

IOP Conf. Series: Materials Science and Engineering 1156 (2021) 012006 doi:10.1088/1757-899X/1156/1/012006

Tiger Barb (Puntiustetrazona) Spawning Using Different Substrates

Rumondang*, **Suhardi Indra Fansyah, Juliwati P Batubara and Khairani Laila**

Study Program of Aquaculture, University of Asahan, Kisaran North Sumatera Indonesia.

* rumondang1802@gmail.com

Abstract. Availability of quality and effective tiger barb seeds is highly limited and leads to unfulfilled market demand. This study was conducted from April - May 2018 in the Laboratory of Aquaculture, Faculty of Agriculture, University of Asahan. Result of analysis of variance (ANOVA) shows that Fstatistic>Ftable (636,097>0,17), thus H0 is rejected and H1 is accepted, hence treatment of different substrates significantly affected egg number of tiger barb. Similarly, the value of fecundity, hatching rate, and survival rate obtained in this study resulted in Fstatistic>Ftable (353.676>0,17) or H0 is rejected and H1 is accepted, hence significant effect on hatching rate was obtained due to substrate treatment and Fstatistic>Ftable (38,741>0,17) H0 is rejected and H1 is accepted which reflected significant influence on tiger barb survival by reason of different substrate treatment. Hatching Rate is obtained by calculating the number of eggs hatched by tiger barbs (Puntiustetrazona). The highest hatching rate of tiger barb in this study was observed in treatment of Hydrillasp (88,14%) followed by palm fiber (72,89%) and water hyacinth (72,75%), while the lowest percentage was obtained by treatment of twigs and raffia rope.

1. Introduction

Aquaculture business, particularly ornamental fish culture, is in great demand by the community for its high potential, such as[1] trade value of ornamental fish which has exceeded the target of 850 million fish. It reached 978 million fish or 115.16 % of initial target and further increased to 1.04 billion fish in 2013. One of ornamental fish originated from Indonesia with high market share, from local to global, is tiger barb (Puntiustetrazona)[2].

Availability of quality and effective tiger barb seeds is highly limited and leads to unfulfilled market demand. Based on observation result, [3] the production of tiger barb seed is not yet optimal even though spawning process of this fish is relatively easy to perform. People normally rely on the activity to catch fish seed in the wild since production aspect of tiger barb seed is still rarely done. Limited number of tiger barb (Puntiustetrazona) culture becomes a problem that requires seriousness in its culture business. The best solution for the development of tiger barb (Puntiustetrazona) culture is possibly done through intensive system. In order to produce high quality seed, many factors are considered, such as good quality broodstock, management of feed, water quality, and spawning substrate as places to attach eggs and as tools to stimulate broodstocks to ovulate[4] [5] [6]. In their habitat, tiger barb (Puntiustetrazona) normally use Hydrillasp as place to attach their eggs. Despite different substrates used by fish farmers in tiger barb culture, information about type of substrate preferred by tiger barb to optimally attach/lay their eggs is still limited. Therefore, the study titled "Tiger Barb (Puntiustetrazona) Spawning Using Different Substrates" was conducted.

2. Methods

The study was conducted from April - May 2018 in the Laboratory of Aquaculture, Faculty of Agriculture, the University of Asahan.

2.1 Hatching Rate/HR

Hatcing rateis defined as the percentage ofhatched eggsto fertilized eggs. Observation was done when hatched eggs were no longer observed. Hatching Rate is determined using the formula:

$$HR = \frac{Numberof\ hatchedeggs}{Numberoffertilizedeggs} \times 100\% \quad (1)$$

2.2 *Survival Rate/SR*

Survival Rate is the percentage of fish still alive which is calculated from stocking until the end of observation. Survival rate of larva in this study is the percentage of number of live larvae during the yolk sac phase:

$$SR = \frac{Numberoffishattheendofobservation}{Numberoffishatthebeginningofobservation} \times 100\% \quad (2)$$

2.3 Fecundity

Gravimetric method is commonly applied to measure fecundity. Its principle is basically similar to volumetric method, yet the volume unit is changed to weight unit. The formula is written as follows:

$$F = \frac{G}{Q} xN \quad (3)$$

Description:
F : Fecundity
G : Body weight (initial weight of fish)/gram
Q : Gonad weight of sample/gram
N : Number of egg of sample/egg

2.4 Data Analysis

Experimental design applied in this study is the Completely Randomized Design (CRD) with 5 treatments and each treatment was replicated 20 times. Analysis was done using the software SPSS ver.21 following the equation below:

$$Yij = \mu + \Pi + \sum ij \quad (4)$$

where:
Yij : Value of observation; *i*=1,2,3...n
μ : General mean
Π : Effect of the-i^{th} treatment; *i*= 1,2,3...n
∑ij : Effect of error due to the-j^{th} replication of the-i^{th} treatment
i : A_0, A_1, A_2, A_3, A_4 (treatment)
j : 0, 1, 2, 3, 4 (replication)

3. Results and Discussion

3.1 Fecundity

Number of eggs produced by fish that are attached and ovulated by tiger barbs (*Puntiustetrazona*) on each treatment substrates of *Hydrillasp*, water hyacinth, raffia rope, twigs, and palm fiber was observed to have different percentage. Ovulation process is the release of egg from its follicle to further enter lumen or abdominal cavity [7][8]. Broodstock of matured tiger barbs (Puntiustetrazona) were used in this study with average weight of 0,27-0,35 gram.

Based on the result of study conducted on tiger barbs (*Puntiustetrazona*), eggs were found mostly attached on palm fiber of approximately 941 eggs, *Hydrilla*sp of 894 eggs, and water hyacinth of 792 eggs, yet no eggs were found attached on raffia rope and twigs. Moreover, the highest hatching rate was obtained by treatment of *Hydrilla*sp (88,14%) followed by water hyacinth (74,56%), and palm fiber (72,89%). Eggs did not hatched on substrate of raffia rope and twigs. Detail result is presented in Figure.

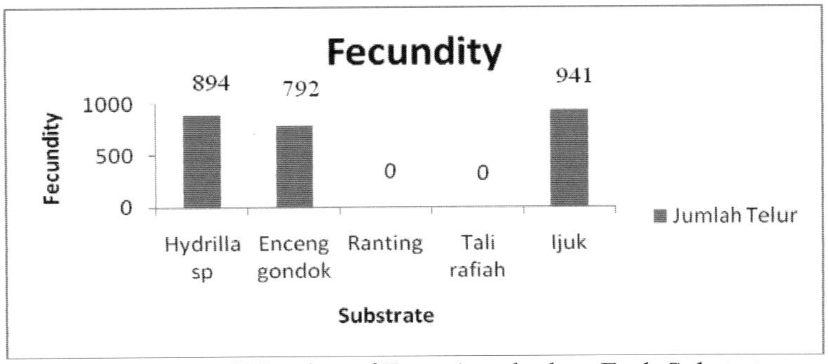

Figure 1. Total Number of Eggs Attached on Each Substrate

Result of analysis of variance (ANOVA) showed that Fstatistic>Ftable (636,097>0,17), thus H_0 is rejected and H_1 is accepted, indicating that different substrate treatment significantly affected tiger barb fecundity, as shown in Appendix 1.

3.2 Hatching Rate
Hatching Rate is obtained by calculating the number of eggs hatched by tiger barbs (Puntiustetrazona). The highest hatching rate of tiger barb in this study was observed in treatment of *Hydrillasp* (88,14%) followed by palm fiber (72,89%) and water hyacinth (72,75%), while the lowest percentage was obtained by treatment of twigs and raffia rope. Detail result is depicted in Figure.

Result of ANOVA (analysis of variance) showed that Fstatistic>Ftable (353.676>0,17) or H_0 is rejected and H_1 is accepted, thus treatment of different substrate significantly influenced hatching rate of tiger barb, as presented in Appendix 2.

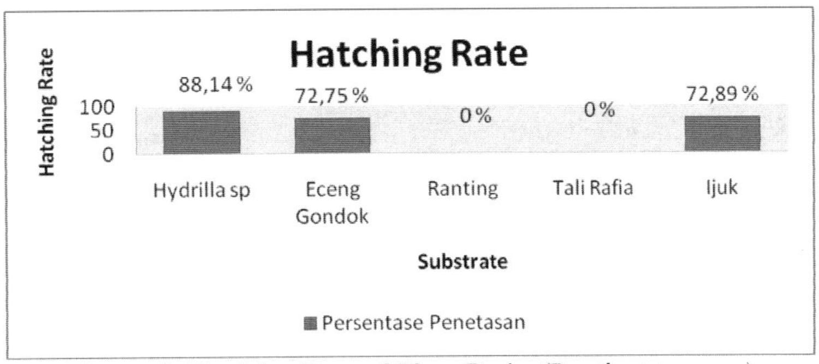

Figure 2. Hatching Rate of Tiger Barbs (Puntiustetrazona*)*

3.3 Survival Rate
Survival Rate is obtained through by calculating the percentage of tiger barb larvae (Puntiustetrazona) that survived the yolk sac phase. Survival rate of larvae in this study is the percentage of larvae observed still alive during the yolk sac phase.

Based on the result of the study, the highest survival rate of tiger barb larvae (Puntiustetrazona) was obtained by treatment of Hydrillasp, followed by palm fiber and water hyacinth. Survival rate of fish larvae in treatment of raffia rope and twigs was zero due to the absence of spawning process done by tiger barbs (Puntiustetrazona). Survival rate of tiger barb larvae (Puntiustetrazona) maintained during yolk sac phase is shown in Figure.

Result of analysis of variance (ANOVA) showed that Fstatistic>Ftable (38,741>0,17) or H_0 is rejected and H_1 is accepted, reflecting significant result on survival rate of tiger barb (Puntiustetrazona) due to different substrate treatment Detail description is found in Appendix 3.

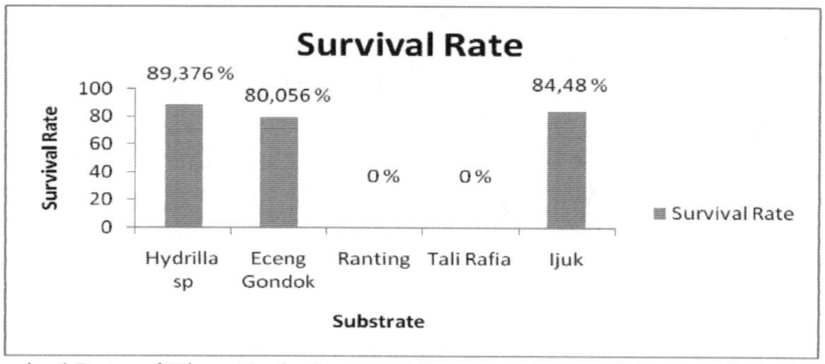

Figure 3. Survival Rate of Tiger Barbs Larvae (*Puntiustetrazona)* during *Yolk Sac* Phase

3.4 Discussion

Based on the result of study conducted on tiger barbs (Puntiustetrazona), eggs were found mostly attached on palm fiber of approximately 941 eggs, Hydrillasp of 894 eggs, and water hyacinth of 792 eggs, yet no eggs were found attached on raffia rope and twigs. Moreover, the highest hatching rate was obtained by treatment of Hydrillasp (88,14%) followed by water hyacinth (74,56%), and palm fiber (72,89%).

Eggs were mostly found in palm fiber since this substrate was put at the bottom of aquarium, thus facilitating fish to attach their adhesive eggs. In addition, fibrous texture of palm fiber also stimulates fish to ovulate. During spawning process, fish will chase one another at the bottom of aquarium or between palm fiber since the substrate was put at the bottom of aquarium, therefore minimizing the chance for eggs to fall into the bottom of aquarium. According to [6][9][10], besides the position of substrate, number of eggs attached on substrate is also affected by sanitation and number of fibrous layer where eggs were attached on, thus more substrate means higher number of eggs attached.

The use of raffia rope and twigs as substrate for fish to attach eggs was found to produce the lowest fecundity since raffia rope is slippery for tiger barbs (Puntiustetrazona) to attach their eggs. Moreover, raffia rope could also bind particles or dirt in the aquarium, thus egg microphyll is possibly entered by other particle, closes, and makes it difficult for sperm to enter. The broodstock of tiger barbs (Puntiustetrazona) were found to be uncomfortable with the presence of synthetic substrate which contains chemical substance. In fact, chemical compound in raffia rope could inhibit the development of embryo during incubation period [11][12][9]. Furthermore, substrate of twigs was observed to have rough and inflexible texture which was found to be unattractive by female tiger barb (Puntiustetrazona) to ovulate since placement of eggs on the substrate is influenced by texture and position of substrate[3][13].

Hatching Rate is determined by calculating the number of eggs hatched by tiger barbs (Puntiustetrazona). The highest hatching rate of tiger barb in this study was observed in treatment of Hydrillasp (88,14%) followed by palm fiber (72,89%) and water hyacinth (72,75%), while the lowest percentage was obtained by treatment of twigs and raffia rope.

It was observed that Hydrillasp plant was extending upwards to the water surface, thus preferred by female fish to attach their eggs. Moreover, concentration of dissolved oxygen in Hydrillasp was higher than that in treatment of palm fiber. [14][15] In fact, Hydrillasp is commonly used by tiger barbs (Puntiustetrazona) to attach their eggs in nature. Even though eggs were mostly found in palm fiber, yet hatching rate of those eggs were lower than that in Hydrillasp since palm fiber is too dense, thus hindering sperms to fertilize the eggs. [9] When the movement of sperms to eggs is hampered, sperms would turn their direction or lose the energy to move which might result in sperm death before fertilizing the eggs.

Survival rate is the comparison between the number of fish at the end of certain rearing period and the number of live fish at the beginning of the period. Eggs will hatch and develop into larvae. The process of larva rearing was done properly. Based on the result of the study, the highest survival rate of tiger barb larvae (Puntiustetrazona) was obtained by treatment of Hydrillasp, followed by palm fiber and water hyacinth, yet survival rate of fish larvae in treatment of raffia rope and twigs was found. Result of ANOVA is presented in Appendix 3.

In this study, larvae were only observed during the yolk sac phase, namely from eggs hatched until the yolk was completely eaten which took around 1-4 days. Based on the observation, the yolk sacs of tiger barbs were not fully disappeared at the same time as they could disappear on day-2, 3 or day-4. Larvae start to adapt and search for foods around them immediately after the yolk sac is completely disappeared [6] .

The high survival rate found in treatment of Hydrillasp has resulted from the high concentration of dissolved oxygen produced by Hydrillasp, thussupplying oxygen in the spawning tank that used Hydrilla sp. Horvath and Peteri (1980) mentioned that dissolved oxygen and temperature greatly affect ovulation. Furthermore, low survival rate in treatment of raffia rope and twigs was caused by different concentrations of dissolved oxygen between treatment of Hydrillasp and raffia rope and twigs.

Hatching rate is also affected by internal factors, such as hormone and yolk volume and external factors, namely temperature, dissolved oxygen, and light intensity [16]. Poor water quality will hinder the growth of embryo, thus facilitating pathogen to attack eggs[3].

Factors of water quality play important role in tiger barb hatching. Spawning tiger barb broodstock observed in this study indicated that water quality during spawning process was still within the range that is suitable for fish production. Excellent environmental conditions for tiger barbs in their tropical nature environment include pH 6-8, DO $\geq$ 5, and temperature 24–28 ^{o}C, while spawning process is optimal at temperature 28 ^{o}C (Verhoef - Verhallen 2000). Water temperature during study ranged between 24–26 ^{O}C, while pH and DO ranged from 6,2–7,8 and 4,10–5,22 ppm, respectively.

4. Conclusions

Based on the result of analysis of variance (ANOVA), Fstatistic>Ftable (636,097>0,17), thus H0 is rejected and H1 is accepted, indicating that different substrate treatment significantly affected tiger barb fecundity. Similar result was observed in hatching rate and survival rate where Fstatistic>Ftable (353.676>0,17) or H0 is rejected and H1 is accepted, showing that treatment of different substrate significantly influenced hatching rate of tiger barb and Fstatistic>Ftable (38,741>0,17) or H0 is rejected and H1 is accepted, reflecting significant result on survival rate of tiger barb due to different substrate treatment.

References

[1] [KKP] Kementerian Kelautan dan Perikanan, "Pedoman Umum Industrialisasi Kelautan Dan Perikanan," J. Chem. Inf. Model., 2012.

[2] E. NOVIANTO, L .y '. 2004.

[3] S. Kirankumar and T. J. Pandian, "Production of androgenetic tiger barb, Puntiustetrazona," Aquaculture, vol. 228, no. 1–4, pp. 37–51, 2003, doi: 10.1016/S0044-8486(03)00132-7.

[4] Z. Arifin and Rumondang, "Pengaruh Pemberian Suplemen Madu Pada Pakan Terhadap Pertumbuhan Dan FCR Ikan Lele Dumbo (Clariasgariepinus)," J. Fish., vol. 1, no. 1, pp. 1–11, 2017.

[5] Rumondang, "Kajian Makanan Ikan Dan Waktu Makan tangkap langsung menggunakan jaring Fakultas Pertanian Universitas Asahan menggunakan buku Illustrations Of Freshwater Plankton Of Japan. Setiap selama penelitian maka dilakukan suhu, kecepatan arus , kecerahan , Inde," in Prosiding seminar nasional multidisplin ilmu, no. November, 2018, pp.

398–407.
[6] A. L. Sinaga and R. J. P. Batubara, "Pengaruh Pemberian Pakan Terhadap Tingkat Kematangan Gonad Ikan Putak (Notopterusnotopterus)," pp. 1–16, 2018.
[7] M. Kobayashi, K. Aida, K. Furukawa, Y. Kok Law, T. Moriwaki, and I. Hanyu, "Development of sensitivity to maturation-inducing steroids in the oocytes of the daily spawning teleost, the kisuSillago japonica," Gen. Comp. Endocrinol., vol. 72, no. 2, pp. 264–271, 1988, doi: 10.1016/0016-6480(88)90209-2.
[8] L. Liu, R. Zhang, X. Wang, H. Zhu, and Z. Tian, "Transcriptome analysis reveals molecular mechanisms responsive to acute cold stress in the tropical stenothermal fish tiger barb (Puntiustetrazona)," BMC Genomics, vol. 21, no. 1, pp. 1–14, 2020, doi: 10.1186/s12864-020-07139-z.
[9] T. Patria, M. Pemijahan, I. Maskoki, D. Menggunakan, B. Substrat, and D. Bakti, "TOMMY PATRIA MARBUN. Pemijahan Ikan Maskoki (."
[10] A. K. Silaban and S. Usman, "Pengaruh Pemberian Pakan Alami Berbeda terhadap Pertumbuhan dan Kelangsungan Hidup Larva IkanNila (Oreochromis Niloticus)," vol. 4, no. 1, pp. 121–138, 2018, [Online]. Available: https://doi.org/10.1016/j.cell.2017.12.025%0A http://www.depkes.go.id/resources/download/info-terkini/hasil-riskesdas-2018.pdf%0A http://www.who.int/about/licensing/.
[11] M. Barik, I. Bhattacharjee, A. Ghosh, and G. Chandra, "Larvivorous potentiality of Puntiustetrazona and Hyphessobryconrosaceus against Culexvishnui subgroup in laboratory and field based bioassay 06 Biological Sciences 0602 Ecology," BMC Res. Notes, vol. 11, no. 1, pp. 1–5, 2018, doi: 10.1186/s13104-018-3902-8.
[12] A. Teknologi, S. Solusi, D. Bidang, and P. Secara, "ISBN : 978-602-72574-5-0 ISBN : 978-602-72574-5-0," no. November, pp. 19–20, 2015.
[13] S. Wargasasmita, J. Biologi, D. Pusat, S. Biodiversitas, and K. Fmipa-Ui, "Ikan Air Tawar Endemik Sumatra Yang Terancam Punah (The freshwater fishes of endemic of Sumatra that threatened species)," Jumal Iktiologi Indonesia., vol. 2, no. 2, pp. 4–49, 2002.
[14] Y. Nagahama, "17α, 20β-Dihydroxy-4-pregnen-3-one: A Teleost Maturation-Inducing Hormone," Development, growth & differentiation, vol. 29, no. l. pp. 1–12, 1987, [Online]. Available: http://onlinelibrary.wiley.com/doi/10.1111/j.1440-169X.1987.00001.x/abstract.
[15] Harniarti, Munir, and M. A. Akib, "Kualitas Jamu Ternak Pada Berbagai Bentuk Sediaan dan Kemasan Prosiding Seminar Nasional 2018 Seminar Nasional Sinergitas Multidisiplin Ilmu Pengetahuan danTeknologi (SMIPT)," Semin. Nas. Sinergitas Multidisiplin Ilmu Pengetahuan danTeknol., vol. 1, no. April, pp. 223–229, 2018.
[16] Rumondang dan E. Paujiah, "Kondisi Plankton Pada Tambak Ikan Kerapu Di Desa Mesjid Lama Kecamatan Talawi Kabupaten Batu Bara, Sumtera Utara," DEPIK J. Ilmu-Ilmu Perairan, Pesisir dan Perikan., vol. 9, no. 1, pp. 107–118, 2019, doi: 10.13170/depik.9.1.14282.

IOP Conf. Series: Materials Science and Engineering 1156 (2021) 012007 doi:10.1088/1757-899X/1156/1/012007

Soil and Plant Nutrients Status of Salak Sidimpuan in South Tapanuli, North Sumatra Indonesia

Yusriani Nasution

Agriculture of Faculty , Universitas Graha Nusantara, Jl. Imam Bonjol 45 Padangsidimpuan, Indonesia.

* yusrianinasution17@gmail.com

Abstract. Salak Sidimpuan is the superior fruit in South Tapanuli. Increased production of salak is very important to meet the needs of consumers in the South Tapanuli area and outside this area. One of the factors that can affect the increase in the quality and quantity of salak sidimpuan plant production is the provision of fertilizers in accordance with the needs of the plant. This research was conducted to determine the need for salak sidimpuan fertilizer. The method used was a survey by measuring soil and plant samples based on land units. The nutrients observed were macro (N, P, K, Ca, Mg and S) and micro (Na, Fe, Cu, Zn, Mn, B and Si) elements. Soil nutrient reserves in salak land, respectively N, P, K, Ca, Mg and S are 46.53 kgha-1, 113.84 kgha-1, 190.52 kgha-1, 183.66 kgha-1, 101 , 47 kgha-1, 69.18 kgha- 1. Thus, the status of soil nutrient reserves in salak land in West Angkola in the macro nutrient group is low. The status of nutrients transported to the harvest in macro nutrients is more than micro nutrients in the sequence K> Ca> N> Mg> P> S. The addition of nutrients to fertilization activities in salak soil is greatly recommended.

1. Introduction

Salak Sidimpuan is a superior fruit in South Tapanuli. Salak Sidimpuan is classified as a palm plant which has a different plavor from other salak fruit. The appearance of the fruit is bigger and sugary, a little sour and this fruit is very popular in the Sidimpuan area [1]. Producing centers in South Tapanuli are in the West Angkola, East Angkola and Marancar Districts. Salak land area in this area is 11,874 hectares with a total production of 340,485 tons / ha [2]. Thus the production of this fruit requirement to be maintained or increased to meet consumer demand both in the city of Padangsidimpuan and to export.

Several factors may affect the increase in the quality and quantity of salak sidimpuan plant production, One of the factors is the application of fertilizers in appropriate with the requirement of the plants. The results of soil analysis become a measurement of nutrient availability that serve the interpretation of the fertilizer requirement assessment. Evaluating the exactitude of nutrient uptake is also determined by the results of the leaf and fruit analysis, as they have an integral part with the soil analysis. This is in accordance with the research of Jiang et al and Shunfeng et al which state that the assessment of soil and leaf nutrient status greatly determines fertilization recommendations [3] [4].

Nitrogen (N), Phospor (P), Potassium (K) are nutrient resources that are competed by plant as limiting factors for plant growth and production factors. Changes in nutrient availability can affect the structure and function of the ecosystem through changes in differences and types [5]. N and K in the soil affect influenced growth of zalacca plants studied by Lestari [6] and Woran et al [7]. Furthermore, Islamy [8] states that the absorption of plant nutrients greatly supports the formation of leaves in zalacca plants. The availability of soil nutrients has an important role in connecting soil and plants in the nutrient cycle mechanism in the biosphere ecosystem.

The diagnosis of soil and plant or fruit nutrients has been studied by: Zhang et al [9]; Shunfeng et al [4] on apples but the diagnosis of nutrient status in zalacca has never been obtained. Increasing the production of salak sidimpuan fruit really requires recommendations for proper fertilization so that information on soil nutrient status, leaves and fruit of salak is needed. This study observed soil nutrient status including macro and micro nutrient elements, namely elements N, P, K, Ca, Mg, S, Fe, Mn, Cu, Zn, B,Mo and Si also observed the status of nutrient content in the leaves and fruit of salak. Soil and leaf sample data were taken for comparison with transported nutrients, and nutrient reserves were available in the soil. Thus this study aims to determine formulation of salak fertilizer requirements.

2. Methods

2.1 Research Location

Soil nutrient status was obtained from the results of soil analysis of the salak center area in West Angkola sub-district, plant tissue analysis obtained from plant samples consisting of leaves and fruit of the salak plant. Sampling was carried out in January- December 2018. Soil and plant tissue samples were analyzed at the Soil Laboratory of Andalas University. Salak leaf samples were taken with the criteria of perfect leaves in the middle of the midrib facing each other. Observation of the elemental content in the leaves is the same as the elemental criteria for the fruit. The nutrient observed were macro (N, P, K, Ca, Mg and S) and micro (Na, Fe, Cu, Zn, Mn, B and Si).

2.2 Soil sample determination and soil analysis

Soil samples were taken based on the provisions of land units by taking the 5 widest homogeneous land units. Determination of soil properties is carried out by various methods, namely: soil pH with the method of determining active aci C-organic with the Walkley & Black method [10], N with the Kjehdal method [11], P with the Olsen method [12], K, Na, Ca and Mg with washing method with NH4OAc 1N pH 7 [13]. Micro nutrient elements (Fe, Cu, Zn, Si, Mn and B) by dissolving method with oxalic acid and measuring with AAS (Atomic Absorption Spectrophotometer).

2.3 Determination of samples and analysis of plant tissue

Salak fruit samples were taken representing land units with the criteria for perfectly ripe fruit while the leaf samples were in the middle position of the midrib facing each other. Observation of the elemental content in the leaves is the same as the elemental criteria for the fruit. The elements observed were macro (N, P, K, Ca, Mg and S) and micro (Na, Fe, Cu, Zn, Mn, B and Si) elements.

2.4 Counting of Harvested Nutrients and Nutrient Reserves

Soil nutrient content from the results of laboratory analysis was tabulated based on the elemental criteria of Hardjowigeno [14]. Harvest transported nutrients are obtained by adding up the nutrients in the fruit and leaves of the salak plant, Meanwhile, nutrient reserves are obtained from the difference between the amount of nutrients transported by the harvest and the nutrients available in the soil. . Elements P, S, Fe, Cu, Zn, Mn, B, and Si (units of ppm) while elements of K, Ca, Mg, and Na (units of meq / 100 g).

3. Results and Discussion

Salak land in West Angkola is the largest area of salak in South Tapanuli, but the yields in this sub-district are based on data from BPS Tapsel [15], the total production of bark is only 8.5 tonnes / ha. Generally, salak farmers in South Tapanuli do not perform intensive fertilization either organic or inorganic fertilizers. The tradition of cultivating zalacca plants is not intensive by not rejuvenating the plants. This is done from generation to generation.

The results of soil analysis in the study resulted in macro nutrient content, namely N, P, K, Ca, Mg, and S which had very low to moderate nutrient status. Micro nutrient content (Na, Fe, Cu, Zn, Mn, B, and Si) has a low to very high nutrient status (figure 1). The status of macro and micro nutrients plays a role in the physiological process of salak plants and can affect the quality and quantity of salak plant production. Soil nutrient status on salak land in West Angkola sub-district can be seen in figure 1.

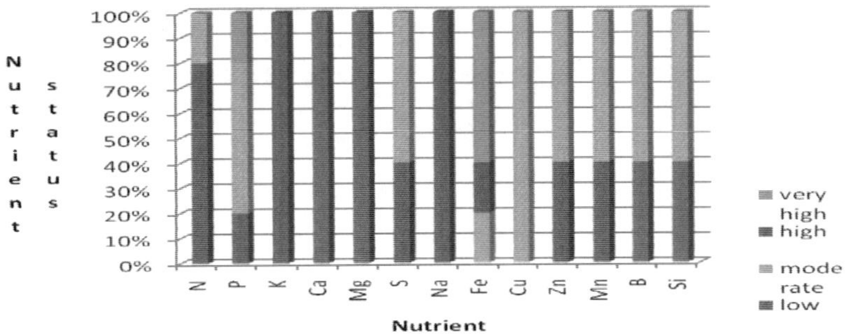

Figure 1. Classification of Soil Nutrient Status of West Salak Angkola Land

The status of soil nutrients in salak land in the macro nutrient group is very low to high, while the micro nutrient status has low to very high nutrient status (figure 1). This indicate that the availability of nutrients in salak land has decreased, due to the absence of input efforts in the form of organic and inorganic materials. In accordance with to Seyyedi et al [16] mineral content in biomass depends on the state of the amount of nutrient availability in the soil. When a nutrient is added through fertilizers to the soil, the amount of nutrients transported to the plant will also increase.

Table 1. Nutrients Transported by Plants (leaves and fruit) West Angkola District

Nutrient	Fruit (%)	Leaf (%)	Amount (%)	kg/ha
N (%)	0.84	1.40	2.24	168.20
P (%)	0.26	0.46	0.71	51.00
K (%)	1.02	1.09	2.11	204.80
Ca (%)	1.00	1.12	2.12	199.20
Mg (%)	0.78	0.87	1.66	156.40
S (%)	0.0014	0.0008	0.0022	0.28
Na (%)	0.0000	0.0001	0.0001	0.01
Fe (%)	0.0008	0.0009	0.0017	0.16
Cu (%)	0.0008	0.0009	0.0017	0.15
Zn (%)	0.0008	0.0010	0.0019	0.16
Mn (%)	0.0011	0.0015	0.0030	0.24
B (%)	0.0004	0.0004	0.0008	0.08
Si (%)	0.0005	0.0005	0.0010	0.10

The amount of nutrients transported by harvest in each element at 15 locations. The highest number of transported nutrients has the sequence, namely, K> Ca> N> Mg> P> S> Mn> Zn> Fe> Cu> B> Si> Na (table1). The amount of nutrients transported in the harvest is based on the production of bark from West Angkola as much as 20 tons / ha [2]. The sequence of the highest number of transported nutrients is K because elemental K, including fertilizer elements, is very important in the growth period of the vegetative and generative phases. This case occur by reason of it directly affects the quantity of salak fruit production. N, P, K, Ca, Mg, and S nutrients must always be maintained in the soil, so that at a certain limit it requires additional fertilizer input as an effort to balance nutrients.

IOP Conf. Series: Materials Science and Engineering 1156 (2021) 012007 doi:10.1088/1757-899X/1156/1/012007

Mehraj et al [17] stated that the content of K, Ca, Fe, P, Mn, and Zn is the mineral content in plants needed by everyday humans. Furthermore, Shabnam et al [18] stated that the availability of elements is influenced by soil acidity so that if the soil is deficient in P, it will limit crop production. Furthermore, deficiency of the micro element Fe can be a problem for agricultural cultivation because plants can experience chlorosis (Xiaoguang et al [19]). Furthermore, Jie et al. [20] stated that the elements transported by plants can be assessed based on the amount transported, namely the more the amount transported indicates that the land is able to produce more biomass because it is supported by the availability of nutrients in the soil. The status of nutrient reserves in salak land can be seen in figure 3.

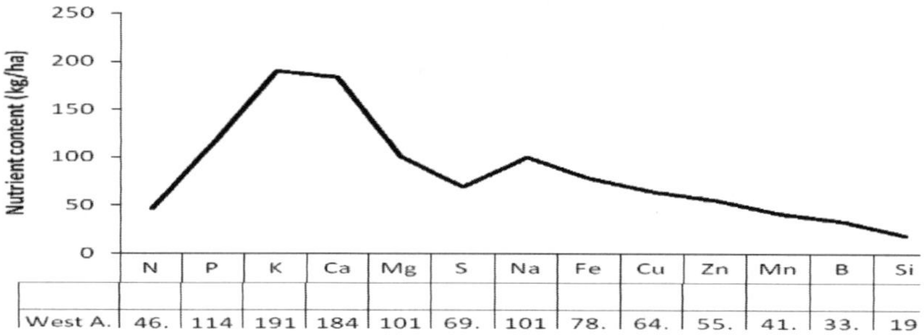

Figure 2. Status of nutrient reserves in the Salak West Angkola area

The amount of soil nutrient reserves in zalacca land after the transportation of nutrients by plants has the highest number of nutrients which are K and Ca elements (figure 2) with low nutrient status according to the criteria of BPT Bogor [21]. Low macro nutrient elements can affect land productivity, thus, low nutrient limitations are not able to produce optimal production without fertilizing the soil. Low nutrient availability is caused by an imbalance in the nutrient cycle at the time of transportation of nutrients transported by harvest, caused by the absence of input to the soil. This is related to the statement by Sumithra et al [22] that the nutrients transported by harvest in plants absorb a lot of nutrients. When this crop is planted continuously without fertilization for 25 years, it can reduce yields. Yuan et al [23] stated that mineral nutrition has an important role in the processes of photosynthesis, respiration and carbohydrate accumulation in plants. Hao et al [24] added that the elements N, P, K, Ca, Mg, and S are taken up by plants in greater amounts than micro nutrients such as Fe and Mn. The need for fertilizer in zalacca land based on nutrient reserves can be seen in table 2.

Table 2. Formulation of salak fertilizer requirements

Nutrient	Amount	Measure	Kg/ha
N	0.20	%	210.73
P	24.89	ppm	263.81
K	0.21	cmol/kg	169.99
Ca	0.19	cmol/kg	163.49
Mg	0.18	cmol/kg	92.70
S	18.40	ppm	234.04
Na	0.17	cmol/kg	85.13
Fe	39.28	ppm	666.13
Cu	32.43	ppm	756.22
Zn	27.92	ppm	118.37
Mn	20.89	ppm	309.96
B	16.76	ppm	106.61
Si	9.49	ppm	80.50

The amount of nutrient reserves in West Angkola after the plants absorb nutrients, namely when the zalacca plants can produce a production of 20 tons / ha as happened in 2012. Based on data from BPS Tapsel [15], the actual production is 8.5 tons, which is potential to be increased to 20 tons / ha / year if nutrients are added in the form of organic and inorganic fertilizers. To maintain and also increase the production of zalacca land, fertilization activities are needed, especially the addition of macro nutrients to the salak land. Thus, the need for fertilizers in the cultivation of salak sidimpuan plants is a must. Shunfeng et al [25] stated that soil and leaf nutrient analysis is an effective diagnostic method for determining nutrient deficiency in plants, fruits, or other plants. This can be recommended to other areas according to soil quality.

4. Conclusions

The status of soil nutrients in salak land in West Angkola sub-district in macro nutrient groups (N, P, K, Ca, Mg and S) is very low to high, while micro nutrients (Na, Fe, Cu, Zn, Mn, B, and Si) have low to very high nutrient status. This shows that the availability of nutrients in zalacca land will continue to decline if there is no effort to add inputs in the form of organic and inorganic materials. The status of transported nutrients from the harvest shows that the macro nutrients transported are more than the micro nutrients, in the order K> Ca> N> Mg> P> S> and the type of nutrient transported the least is Na. The nutrient reserves in the soil have a very low nutrient status. The addition of nutrients in fertilization activities is highly recommended so that the quality and quantity of salak sisimpuan production can be increased.

References

[1] Nasution, Y. A Rasyidin., Yulnafatmawita and A Saidi. "Rainfall Interception of Salak Plants (Salacca sumatrana Becc) in North Sumatera Indonesia". International Journal of Advanced and Innovative Research (2278-7844)/ Volume 7 Issue 7, 2018.

[2] Tapanuli Selatan. South Tapanuli in Numbers. Agency for Extension Agriculture and Food Security. 62 p, 2015.

[3] Jiang, YM., Zhang, HY., Zhang, FS. The Northem deciduous Fruit Threes Nutrient Resource Integrated Management Theory and Practice. China Agricultural University Pess. Beijing :p 27-30, 2007. (in Chinese).

[4] Shunfeng, GE., Z Zhanling., P Ling., C Qian dan J Yuanmao. "Soil Nutrient Status and Leaf Nutrient Diagnosis in The Main Apple Producing Regions in China". Horticultural Plant Journal. 4 (3) : p. 89 -93, 2017.

[5] Tian, L., L Zhao., X Wu., G Hu., H Fang., Y Zhao., Y Sheng., J Chen., J Wu., W Li., CL Ping., Q Pang., Y Liu., W She., T Wu dan X Zhang.. "Variations in Soil Nutrient Availability Across Tibetan Grassland From The 1980s To 2010s". Geoderma. 338 p. 197-205, 2017.

[6] Lestari, R, Georg E dan Susanne H-K. "Fruit Quality Changes Of Salak "Pondoh" Fruits (Salacca Zalacca (Gaertn.) Voss) During Maturation And Ripening". Journal Of Food Research. ISSN 1927 – 0887 E ISSN 1927 – 0895. Published By Canadian Center Of Science And Education. Vol 2. No. : 1, 2013.

[7] Woran, RF., R Nangoi., Jeanne E. Lengkong. The Study of Physical and Land Chemical Properties on The Green Plant Area (Salacca zalacca) in Pangu Village District Minahasa Tenggara. Cocos. Vol.1 No.1, 2018.

[8] Islamy, D. Identification of Soil Nutrient Characteristics and Plant Nutrient Content Associated with the Local Salak Flavor of Sumedang. Major in Land Resource Management, Faculty of Agriculture Institut Pertanian Bogor. 39 p, 2010.

[9] Zhang, Q., Wei QP., Jiang RS., Liu XD., Liu HP., Wang XW. CorelationAnalysis of Fruit Mineral Nutrition Contens with Several Key quality indicators in "Fuji" Apple. Acta Hortic Sin, 38: 1963-1968, 2011. (in Chinese).

[10] Schumacher, B A. "Methods For The Determination Of Total Organic Carbon (TOC) In Soils and Sediments". United States Environmental Protection Agency Environmental Sciences

Division National Exposure Research Laboratory. p. 25, 2002.

[11] Bremmer, JM dan CS Mulvaney. "Nitrogen- Total".in A.L. Page (Editor) .Method of Soil Analysis. Part 2. Madison,Wisconsin; American Society of Agronomy&Soil Science Society of America, 1982.

[12] Olsen, SR dan LE Sommers. "Phoshorus", dalam A.L. Page (Editor) .Method of Soil Analysis. Part 2. Madison,Wisconsin; American Society of Agronomy&Soil Science Society of America, 1982.

[13] Thomas, G W. "Exchangeable Cations". dalam A.L. Page (Editor) .Method of Soil Analysis. Part 2. Madison,Wisconsin; American Society of Agronomy&Soil Science Society of America. P.3-56, 1982.

[14] Hardjowigeno, S. Soil science., Jakarta: Akademika Pressindo, 1995.

[15] BPS Tapanuli Selatan (Badan Pusat Statistik Tapanuli Selatan). 2018. West Angkola District in Numbers 2018. Nomor 12030.1715. ISBN. 978-602-1255-50-6. Published by BPS Kabupaten Tapanuli Selatan.

[16] Seyyedi, SM., PR Moghaddam dan MN Mahallati. "Weed Competition Periods Affect Grain Yield and Nutrient Uptake of Black Seed (Nigella Sativa L.)". Horticultural Plant Journal. 2 (3): Hal. 172–180, 2016.

[17] Mehraj, H., Y Nishimura dan K Shimasaki. "Analysis of Essential Macro-micro Mineral Content of Twelve Hosta Taxa". Annals of Agricultural Science. 62, 71-74, 2017.

[18] Shabnam, S., AH Khoshgoftarmanesh., M Afyuni., M Shrivani dan R Schulin. "The Effect of Preceding Crop on Wheat Grain Zinc Concentration and Its Relationship to Total Amino Acids and Dissolved Organic Carbon in Rhizosphere Soil Solution". Biology and Fertility of Soils. Vol. 50. Issue 2. p. 239-247, 2018.

[19] Xiaoguang, LI., G Ying., Y Shuai., Z Jianghomg dan Z Jianguang. "Characterization and ExpressionAnalysis of PbCS1 Gene Regulated by Iron", 2016.

[20] Jie, Y., Z Haijin., C Xiaoan dan S Le. "Effects of Tillage Practices on Nutrient Loss and Soybean Growth in Redsoil Slope Farmland". International Soil and Water Conservation Research. Vol. 1 No.3. Hal. 49-55, 2013.

[21] BPT (Balai Penelitian Tanah) Bogor. "Characteristics, Potential, and Technology of Ultisol Soil Management for the Development of Dryland Agriculture in Indonesia ".Jurnal Litbang Pertanian Vol. 25 (2). 2006.

[22] Sumithra, R., M Thushyanthy dan T Srivarathasaran. "Assessment of Soil Loss and Nutrient Depletion Due to Cassava Harvesting: A Case Study from Low Input Traditional Agriculture". Soil and Water Conservation research. Vol. 1. No.2. p. 72-79, 2013.

[23] Yuan, H., Z Liqiang., K Quisheng., C Fei., N Mengliang., X Junjun., MA Nawaz dan B Zhilong. "Comprehensive Mineral Nutrition Analysis of Watermelon Grafted ontoTwo Different Rootstocks". Horticultural Plant Journal. 2 (2): 105-113, 2016.

[24] Hao, W., M Abinaya., C Yuze dan JB Ryong. "Effect of Different Cultivation System on The Accumulation of Nutrient and Phytochemicals in Ligularia Fischeri". Horticultural Plant Journal. Vol.4 (1) : p. 24-29, 2018.

[25] Shunfeng, GE., Z Zhanling., P Ling., C Qian dan J Yuanmao. "Soil Nutrient Status and Leaf Nutrient Diagnosis in The Main Apple Producing Regions in China". Horticultural Plant Journal. 4 (3) : p. 89 -93, 2017.

Acknowledgments

This work was supported by Graha Nusantara University. I also would like to thank the editor and anonymous reviewer for their constructive comment on the manuscript.

BIM Implementation in Public Construction Projects in Indonesia

Nurkhasanah Rina Puspita* **and Fithriyah Patriotika**

Faculty of Engineering, University of Graha Nusantara

* n.r.puspita@outlook.com

Abstract. The delayed of BIM advancement in Indonesia has proffered that Indonesian government awareness to generate an optimum knowledge and research in BIM implementation is frequently low. In order to catalyst some infrastructure projects which based government projects, the contractor companies should put their creativity and their improved knowledge to win the government projects. Regarding the existing challenges that frequently exist in several projects, BIM has been appeared as a sophisticated technology- based solution to overcome them. However, the level of BIM development in Indonesia nowadays has not been confirmed yet due to the BIM research evolvement is slowly. Even though, there are some companies which has utilised BIM either in their companies or their project such as the construction of Bakauheni - Terbanggi Besar toll road, the construction of Solo - Kertosolo toll road, etc. Nevertheless, the effectiveness and efficiency of BIM utilisation in a contractor company need to measure as an attempting to assist BIM research evaluation in Indonesia. By the questionnaire survey, this research is going to find out how BIM take the responsibility to reduce the complexity of diversity project. Moreover, the benefits and drawbacks which BIM contributes for several projects which have already through BIM's role need to assembly as a further consideration for Indonesian government to arrange the convenience regulation regards to adopt BIM as the obligations later. Additionally, the assessment of the stakeholder's role in BIM also necessary to evaluate since the optimum of BIM's assignment in construction industry involving not only the contractor but also the consultant and the owner. Thus, the research is expected to endowment the further knowledge for BIM implementation in Indonesia.

1. Introduction

Regarding the construction industry development, Indonesia as one of the developing countries in the world particularly in Asia is sluggish for supporting the technology developing and the technology adopting by its competitor countries such as Singapore. Additionally, the issues which frequently appear in construction site still concern on delayed problem. For example, there are not few projects in Indonesia which still witness with the delayed schedule. It is caused by improper plan which arranged since the early stage of life-cycle project beside natural factors such as weather, disaster, etc.

In a further research, the background factor of lately development the construction industry in Indonesia is caused by the weakness of government attention towards improving the regulation of the renewable and sophisticated are carried out by the small-scale and the medium-scale construction companies in order to realisation the infrastructure equalisation in Indonesia. Indirectly, it has mean that the role of these two types company has been considered by government. Therefore, the awareness regarding capability of these companies have to be a main point which must to be arranged concretely by government in order to reduce the existing challenges occur in a project.

Delayed work, excessed cost, and poor of work's quality have still happened frequently either in a big or small project, particularly in government projects. Unsurprisingly, some projects which sponsored by government mostly face some technically issues. Nowadays, the existing challenges which usually occur in a project can be overcome by the current technology in construction industry which named Building Information Modelling (BIM).

Generally, BIM's concepts have been developing since the earliest 1960s [5] [4] stated BIM is a set of building modelling processes that based technology and involves construction technology. On the other hand, Indonesian government targeting that the level of infrastructure developing must to significant increase. Additionally, [8] states that 93% of Indonesian government projects digital system to improve accuracy in design stage. Furthermore, [3] a chain of technology that will increase productivity whilst rectifying design, construction practice and practice maintenance which aims to enhance collaboration inter-organisation in construction industry. Additionally, [1] BIM is understood as the approaching current tools which described a variety of activities in object-oriented Computer Aided-Design (CAD).

On the other hand, the development of BIM is noticed as the sophisticated technology which has the complete solution towards the challenges occasion in construction industry. In the other words, BIM has been providing interoperability capacities and more intelligence [1]. To conclude, BIM is not only a process but also a viability technology which can summarize the set of processes which prevails in construction industry, such as reduce risk value, improving organisation management, etc.

1.1 BIM Development in Indonesia

As knowledge information, BIM have been used by some companies in Indonesia since 2012 [6]. However, despite the awareness level of BIM knowledge is high around 70%, the level of BIM implementation is still low which has 38% [2]. In theory, the advancement of BIM research in Indonesia is not mature yet. The reason behind, it is indicated by poor of publishing international journal regarding BIM implementation in Indonesia [6].

On the other hand, during the improving of construction industry necessity and the level of competition among the contractor companies in Indonesia, the graphic of BIM development in Indonesia has witnessed significant increased. Based on some reports which have been published in either the website or online news, there are some project examples which successfully implement BIM in the projects particularly Indonesian government projects. For instance, the construction of port in Sibolga, the construction of Indonesian Bank in Gorontalo, etc [7].

Despite BIM's projects in Indonesia have necessary increased, the regulation which arrange the BIM implementation in Indonesia has not organised yet. Nowadays, the company-scale's which has been applied BIM as the current technology is only the large-scale company. However, BIM is only utilised as the supporting technology by some, and others use BIM is only for project.

2. Methods

It is important to describe systematically that to support this research both qualitative approach and quantitative approach are the most appropriate methodology to choose for the research. The reason because these methods are the scientific method in which the problems discussed in this study must stand singularly and freestanding from objective reality as well as the experience behind from the respondents to answer the question is needed to enrich the answer of questionnaire. The mixed approach is assumed to be so repeatable, and able to isolate from the uncompromising reality of the cause and effect under investigation. Consequently, these approaches are appropriate because it tends for theory testing rather than theory building.

Regarding to Saunder's research onion, the methodology approach which will use for this research is:

1. Philosophy: Positivism. The reason is the objectives point of view needed as the primary data.

2. Approach: Deductive. Deductive research is leading to use the quantitative methods which is chosen as the methodology for the research.
3. Strategy: Questionnaire Survey
   a. Survey's conduct in this study aims to separate the object which understood by the author with the results of the survey filled by the respondent. This is expected to get independent result without mixed with the author's point of view.
   b. The survey results are premier data to obtain the final result. The survey will be conducted on 3 types of construction company level consisting of 2 small-sized firms, 2 medium-sized firms and 4 large- sized firms.
   c. The survey results are premier data that will be managed systematically to obtain the final result. The survey will be conducted on 3 types of construction company level consisting of 2 small-sized firms, 2 medium-sized firms and 4 large- sized firms.
   d. Respondents will be sent in the form of questionnaires via email in the form of questions related to the readiness of the Indonesian government in the process of standardizing the implementation of BIM and understanding of BIM application procedures in a structured.
   e. Due to the limited time and distance in the study, the respondents are the construction company consisting of 2 small-sized firms, 2 medium-sized firms in the city of Padangsidimpuan and 4 large-sized firms in Jakarta will be sent emails in the form of questionnaires to their respective companies. If the respondent did not answer the survey via email, then the survey will be continued through structured interviews to responded either via email or skype.

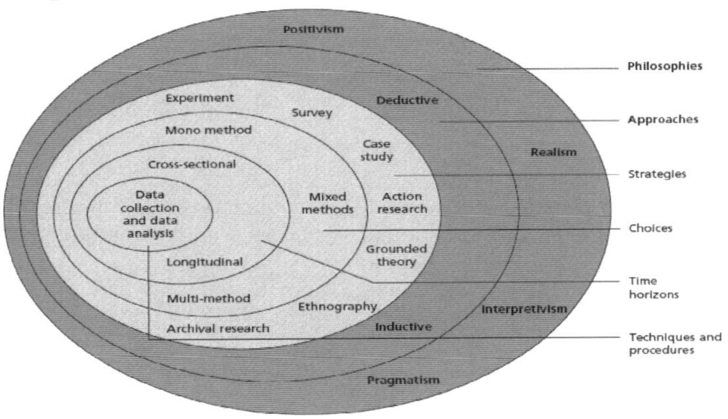

Figure 1. Research Onion

2.1 Problem and Solution

2.1.1 Problem

1. Limitations of language will be a major obstacle because the survey will be conducted on respondents in Indonesia.
2. Time difference between respondents and authors.
3. Understanding of questions asked to respondents

2.1.2 Solution

1. Questionnaires will be sent in two languages where for respondents who can fill in accordance with the language skills of each.
2. The author will follow the work time of respondents in the survey period.
3. The author will double-check the respondent's answer if discrepancies are found in the answer to the questionnaire.

3. Results and Discussion

After the respondents had filled out the questionnaire, the result of the first question in the questionnaire will be elaborated. The table below shows the response which gathered from the questionnaire survey.

3.1 Q1: What type is your company?

In the first main question, the respondents are provided two answers which are "Small" and "Large". The respondent can choose one of the answers which according to their company scale. In order to analyse the data in SPSS, the answers are numbered by the researcher who "Small" is number "1" and "Large" is number 2.

3.2 Q2: Does your company familiar with BIM (Building Information Modelling)?

To generalisation the answer from the respondents, the researcher put two answers due to aid in analysing the data which gathered from the survey. The answer number "1" has meaning "YES" while for the answer "NO" the researcher symbolized with number "2".

3.3 Does your company have been implemented BIM (Building Information Modelling) in your company organisation?

There are two indicators answer to question number 3 where the answer number 1 has meaning "YES" while the answer number 2 is for "NO".

3.4 Q4: Is BIM (Building Information Modelling) your owner requirement before the project started?

The third question provides two options which respondents should choose as their answer. The first answer is signed as number "1" which has meaning "YES" and the second answer is "NO" which also signed as number "2".

3.5 Q5: Does BIM (Building Information Modelling) have been applied in your project?

The question in number five is presented by two answers which in the first answer is symbolized with number "1" and the second answer with number "2". Both answers have meaning "YES" and "NO" respectively.

3.6 Q6: Which percentage is BIM (Building Information Modelling) has been used in your project?

In this question, the researcher gave an option to the projects which chose "No" in the previous question to finish the survey because the next questions are only for the respondents who have implemented BIM in their projects. Furthermore, in order to levelling answer from the respondents, the researcher established the group answer to choose.

Here by the answer which the respondents can choose:
0 - 20% = 1
21 - 40% = 2
41 - 60% = 3
61 - 80% = 4
81 - 100% = 5

3.7 Q7: What is BIM's (Building Information Modelling) advantageous in your project?

In the research, the respondent which involved BIM in its projects have to choose the options of what BIM's positive impacts for their effectiveness in construction progresses. The respondent is pleased to choose more than one option. Figure 2 presents the result of respondent options toward BIM's advantageous. From the bar chart, it can be obtained that better communication and coordination have the highest option. Regarding the detail of the respondent response, the table below shows which companies that are using BIM in their project or not.

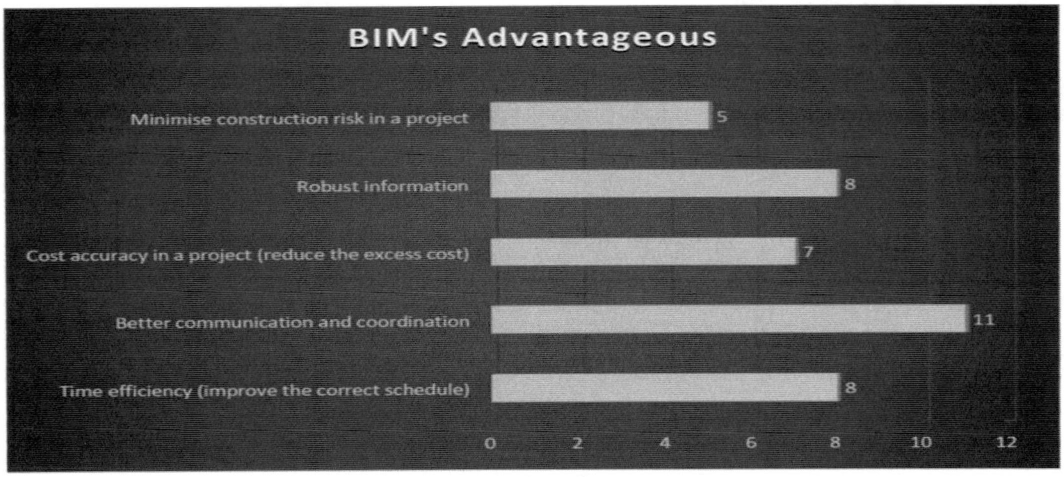

Figure 2. BIM's Advantageous

From the figure 2, there are only six companies which totally number of projects that applied BIM in the projects are seventeen. Half of the respondents are the active user in BIM while half others are not the user of BIM.

3.8 Q8: What is BIM's (Building Information Modelling) barriers to adopt either in your project or your company organisation?

In the questionnaire answers, the respondents are offered to choose more than one answer. The question is only filled by the respondent which its project witnesses with BIM. Figure 3 illustrates that lack of the resources in BIM implementation is the highest number that respondents choose. It can be concluded that necessity of BIM resources is still high among construction companies in Indonesia. However, the resistance to change has the lowest weakness in BIM regarding the improvement of management organisation in company is easier than finding the expert of BIM.

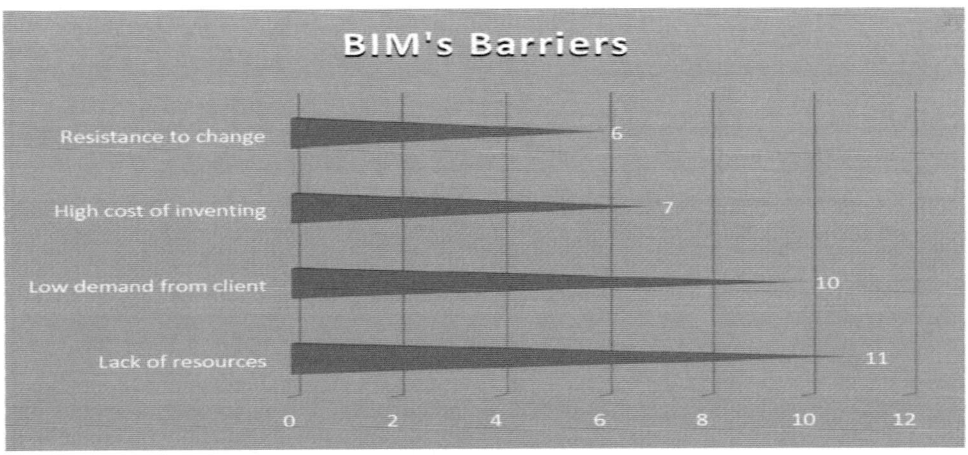

Figure 3. BIM's barriers

3.9 Q9: Which stage is BIM (Building Information Modelling) utilised in your project?

In the research, the question regarding the usage of BIM in life-cycle project is provided with three answers which are pre- construction, construction and post- construction. The respondent is pleased to choose answer that relate in its own project.

3.10 Q10: Which stakeholder does involve in BIM (Building Information Modelling) usage in your project?

By the last question in the survey, the parties which have been entangled in BIM utilising can be measured. The respondent is prepared three options which can be chosen as what its project witness is. After comparing and analysing the methodology which convenience and fit with the condition of research environment, the researcher decided to do a questionnaire survey. Since the object of the study particularly mentioned is Indonesia, then the author arranged the sample size and the respondents targeted. Thereby, there were selected 30 respondents which would be the participants of the research.

Additionally, they are an engineering staff in the different projects from three the state- owned construction company as the large- scale company in Indonesia and five construction local companies in Padangsidimpuan, a small town in Indonesia, as the small-scale companies.

4. Conclusions

Surprisingly, the result of the survey was pleased where the number of respondents witnessed a significant increase. There were 35 respondents from twelve different companies in the research, hence, the data which had been analysed more than the expectation. The data analysing used SPSS and Microsoft Excel to gain both the statistic and the percentage result.

References

[1] Ghaffarianhoseini, A. et al., Building Information Modelling (BIM) uptake: Clear benefits, understanding its implementation, risks and challenges. Renewable and Sustainable Energy Reviews, Volume 75, p. 1046–1053, 2017.

[2] Hanifah, Y. Awareness dan Pemanfaatan BIM: Studi Eksplorasi. Temu Ilmiah IPLBI 2016, pp. 49-54, 2016.

[3] Miettinen, R. & Paavola, S. Beyond the BIM utopia: Approaches to the development and implementation of building information modelling. Automation in Construction, Volume 43, pp. 84-91, 2014.

[4] Salman, n.d. Penggunaan BIM Pada Proyek Infrastruktur, s.l.: s.n.

[5] Smith, P. BIM implementation - Global strategies. Procedia Engineering, Volume 85, pp. 482-492, 2014.

[6] Telaga, A. S. A review of BIM (Building Information Modeling) implementation in Indonesia construction industry. IOP Conference Series: Material Science and Engineering, Volume 352, p 012030, 2018.

[7] Pramudiptha, K. TEKNOLOGI KONSTRUKSI Agenda: Apa itu BIM Mengapa menggunakan BIM Kapan menggunakan BIM Siapa saja pengguna BIM Bagaimana proses BIM, 2018.

[8] Wasito, F. independensi.com. [Online] Available at: independensi.com/2018/02/01/kontraktor-kecil-dan-menengah-kerjakan-93-paket-di- kementerian-pupr/ [Accessed 09 09 2018].

The Development of a Software Tool for Improvement Tracking System Using Dynamic Systems Development Methodology

Ashari Imamuddin*, Mohamand Anas Sobarnas and Iskandar

Faculty of IT, Sekolah Tinggi Teknologi Muhammadiyah Cileungsi (STTMC), Indonesia.

* ashari@sttmcileungsi.ac.id

Abstract. The present study aims to develop a software tool for improvement tracking system for a waste management company in Indonesia. The tool was built as a system to track the improvement process of following up findings of the validation and verification process in the management system of the company. The company also obtains a few audits by the authorities, certification bodies, and its customers. The company applied several management systems in assuring all persons in charges are responsible for following up on the improvement findings. The outcome of the study is a proven software tool that has supported the top management in making decisions in terms of the improvement processes. The system automatically reports to the top management when the findings of each individual responsible person have not been executed yet as well as the top management can take direct decisions and actions to the current status of their improvements processes.

1. Introduction

A configurable business process model (sometimes referred to as a reference business process model) may be configured to meet the specific requirements of an organization[1]. As information technology today is strongly intermeshed with business processes, requirement specifications for IT-implementations obtain much more significance for enterprises' prosperity. The on-going technological progress precipitates more valuable IT. This facilitates enterprises to offer more efficient and convenient processes to the customer[2]. During the determination of the appropriate measures of the process models we focused primarily on the BPMN notation. This notation is currently very well sophisticated, it can be considered as a standard tool in the modelling of the business processes in a wide group of process-controlled organizations[3]. The system supports the company's mission in terms of preserving its commitment to minimize its customer's environmental risk[4].

This research is a software modeling for Improvement Tracking System (ITS) at a waste management company which includes requirements specification, process, data, and interface modeling, and application architecture. The system is managed and utilized by QEHS (quality, environmental, health, and safety) department of the company. The development process includes the stages of system analysis, system design, programming, testing, and installation[5],[6]. There are the software capabilities that must be present in order for the user to carry out the services provided by the feature, or to execute the use case.

To develop the improvement tracking system using agile methodology produces higher quality software in a shorter period of time. Agile methodologies were developed to streamline the development process and remove barriers to accepting business requirement changes during the development process[7], agile software development framework to assist software prac-titioners in developing small and medium sized projects, the success of which is pointed out in several industry cases [8] same as software Improvement Tracking System (ITS).

The project is to develop for improvement and customizing the existing ITS. It covers project scope, product description, system features, system interface, functional requirements, and non-functional requirements. The development process includes the stages of system analysis, system design, programming, testing, and installation. For the redesain business process modeling the develop this software use the Agile method with Dynamic Systems Development Methodology (DSDM)[9], because the method of the development process involves the user and matches the project to be worked on.

2. Methods

There are several methodologies of Agile encompasses various methodologies including: Adaptive Software Development (ASD), Agile Unified Process (AUP), Crystal Methods, Dynamic Systems Development Methodology (DSDM), eXtreme Program-ming (XP),Feature Driven Development (FDD), Kanban, Lean Software Development, Scrum and Scrumban[10]. In this develop software ITS, use Agile method of Dynamic Systems Development Methodology (DSDM) illustrated in figure 1, which has of five phases [11] , As the Feasibility Study, Business Study, Functional Model Iteration, Design and Build Iteration and Implementation.

3. Results and Discussion

The implementation of the develop software ITS in waste management in company will use the Agile - DSDM method as below stage:

Mainly the system supports one of the processes of QEHS Department of company as figure 1. The system is started by initiating finding by QEHS Admin. The system is accompanied by features of system notification by email as figure 2.

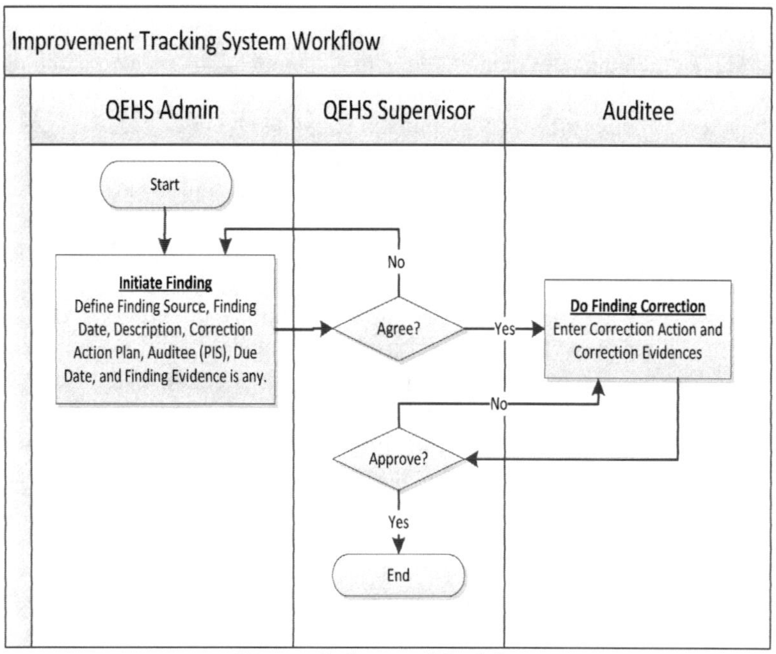

Figure 1. System Workflow

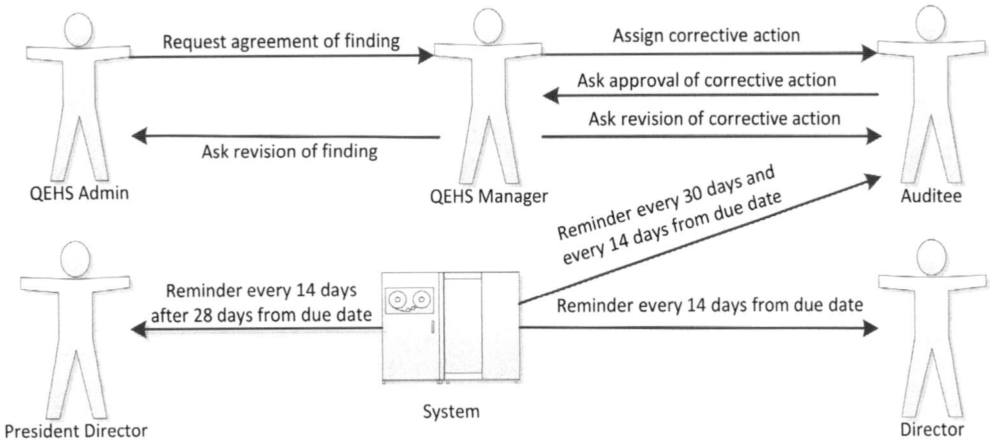

Figure 2. System Notification

Major features of the system are finding (recording, tracking, and approval), master/lookup tables, application security, email notification, and dashboard. As the conceptual database design using ER (entity relationship) model same as figure 3:

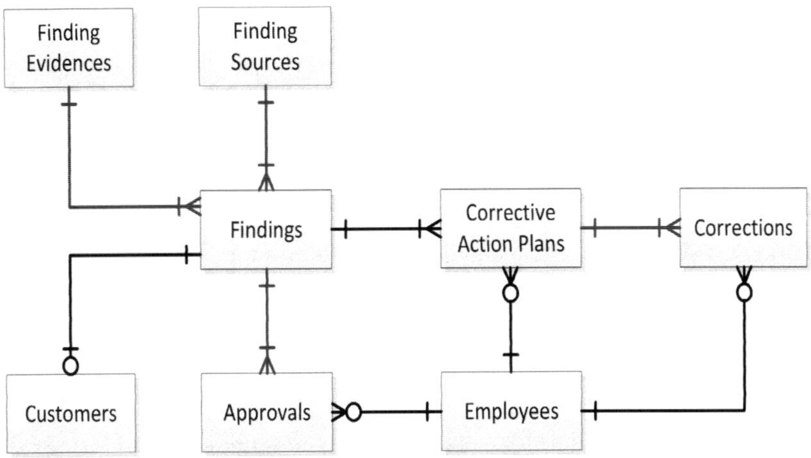

Figure 3. Entity relationship diagram

The database is located into two physical servers. Most entities are placed at ITS database on company SQL Server except Customers table at the company's existing Operation database on the same server as figure 4.

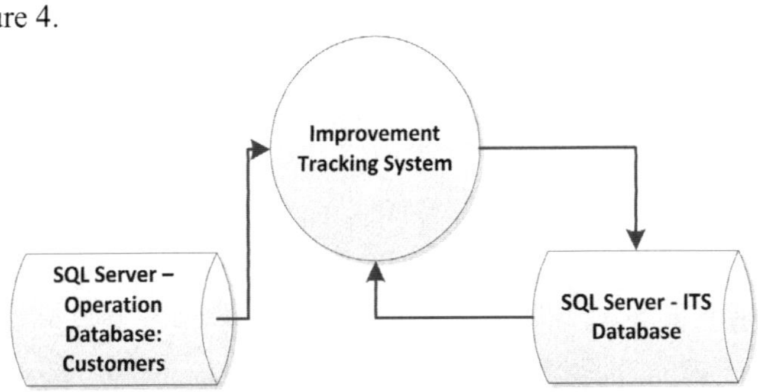

Figure 4. Conceptual Database Design

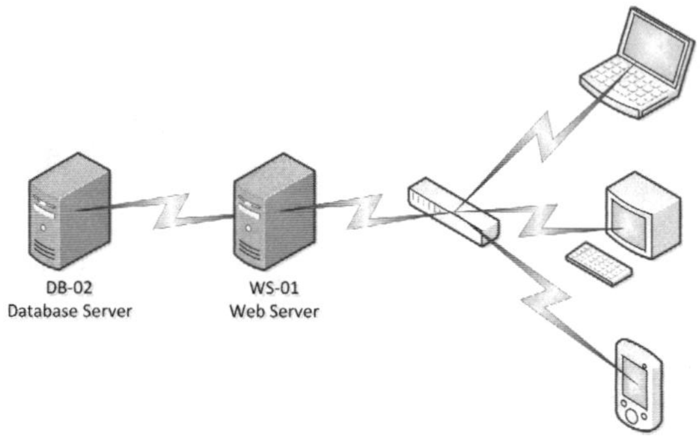

Figure 5. Application Architecture

Table 1. Description of Function

Requirement Code	Function Name	Description	Actor
ITS-SRS-001	Create Finding	Creating new or revising finding and define corrective action plan (CAP).	QEHS Admin
ITS-SRS-002	Open Finding	Opening a finding to assign CAP to responsible department PIC.	QEHS Manager
ITS-SRS-003	Assigned Finding	Entering corrective action on the assigned CAP.	Responsible Department's PIC
ITS-SRS-004	CA Approval Finding	Approving or rejecting correction of assigned CAP by Responsible Department PIC.	QEHS Manager
ITS-SRS-005	Closed Finding	Viewing close correction of assigned CAP by responsible department PIC.	All Users
ITS-SRS-006	Inquiry Finding	Searching for any statuses.	All Users
ITS-SRS-007	Title	Entering, modifying or viewing job titles for employees.	QEHS Admin
ITS-SRS-008	Departments	Entering, modifying or viewing department for employees.	QEHS Admin
ITS-SRS-009	Employees	Entering, modifying or viewing employees and his/her title and department.	QEHS Admin
ITS-SRS-010	Finding Sources	Entering, modifying or viewing finding source.	QEHS Admin
ITS-SRS-011	User Accounts	Entering, modifying or viewing user account and defining its user access role and accessible finding sources.	System Administrator
ITS-SRS-012	User Access Roles	Entering, modifying or viewing a user access roles and defining the accessible menu of the user role.	System Administrator
ITS-SRS-013	Dashboard	Show status report and performance report.	All users

In order to address system requirements we also apply application security, system email notification, and dashboard system.

1. **Application Security**: is to ensure users who access the system are only defined users in the system with a particular role and they only access the system as the defined role. One user only uses one role.

2. **Email Notification**: is triggered either by user event or automatic by system based on correction due date.
   - User event notification works when QEHS Admin submits a finding then the system notifies QEHS Manager for his/her agreement of the finding.
   - System event notification works when a corrective action plan is not done: The system notifies the responsible department PIC and his/her superior every 30 days before the due date.
   - The system notifies the responsible department PIC, his/her superior, his/her director, and the President Director every 14 days from 28 days after the due date.
3. **Dashboard**: is to inform users of performance of following up findings in statistic reports.
   - Finding & Corrective Action Status: shows number of finding for each step: Create, Open, Assigned, In Progress, CA Approval, and Close.
   - Due Date Performance: shows number of finding correction closed before the due date, at due date, and after the due date.
   - Remaining Findings vs. Remaining Due Date Findings: shows trend of number of Remaining Findings compares against Remaining Due Date Findings.

This stage describes detailed functional requirements associates to the features. All SRS of functional requirements below are described in detail Use Case Specifications in table 1 as below.

In this stage as describe of details requirements from the comments users as Simplification workflow, Modifying Finding page, Adding Finding Category in Finding Sources, Applying email notification, Removing Work Order module, Modifying on user role module and Dashboard.

The final stage of the application ITS after modeling the business process is implementing the design into software application by constructing design of database, process, and interface into a working software prototype as ITS tool.

4. Conclusions

As the existing workflow in figure 1 about modeling for software Improvement Tracking System (ITS) as software tool of a system to track improvement process of following up finding of validation and verification process in management system of the company also obtains a few audits by the authorities and its customers. The application as ITS tool complied and met user and business requirements. After testing for modeling business process where in the main function in the improvement process of this business process as which expects better response time improvements and with full control from high management levels so that it can be expected by management in proper waste management.

References

[1] N. M. Bobrovskij, D. G. Levashkin, I. N. Bobrovskij, P. A. Melnikov, and A. A. Lukyanov, “The Modelling of Basing Holes Machining of Automatically Replaceable Cubical Units for Reconfigurable Manufacturing Systems with Low-Waste Production,” IOP Conf. Ser. Earth Environ. Sci., vol. 50, no. 1, pp. 0–8, 2017.

[2] T. V. Bobrova, Y. V. Kodentseva, and M. S. Perfiliev, “Parameterization of environmental components in the information-modeling context of roads,” IOP Conf. Ser. Mater. Sci. Eng., vol. 698, no. 6, pp. 0–6, 2019.

[3] H. Gomaa, L. Kerschberg, and G. A. Farrukh, “Domain modeling of software process models,” Proc. IEEE Int. Conf. Eng. Complex Comput. Syst. ICECCS, pp. 50–60, 2000.

[4] S. Banerjee, X. (Simba) Chang, K. Fu, and G. Wong, “Corporate Environmental Risk and the Customer-Supplier Relationship,” SSRN Electron. J., 2014.

[5] V. Ciaschini, M. Canaparo, E. Ronchieri, and D. Salomoni, “Evaluating predictive models of software quality,” J. Phys. Conf. Ser., vol. 513, no. TRACK 5, pp. 0–7, 2014.

[6] G. Kumar and P. K. Bhatia, “Comparative analysis of software engineering models from traditional to modern methodologies,” Int. Conf. Adv. Comput. Commun. Technol. ACCT, pp. 189–196, 2014.

[7] J. A. Livermore, “Factors that Impact Implementing an Agile Software Development Methodology,” pp. 82–86, 2007.
[8] S. Lee and H. Yong, “Agile Software Development Framework in a Small Project Environment,” vol. 9, no. 1, pp. 69–88, 2013.
[9] M. Coram and S. Bohner, “The impact of agile methods on software project management,” Proc. - 12th IEEE Int. Conf. Work. Eng. Comput. Syst. ECS 2005, no. March, pp. 363–370, 2005.
[10] M. Al-Zewairi, M. Biltawi, W. Etaiwi, and A. Shaout, “Agile Software Development Methodologies: Survey of Surveys,” J. Comput. Commun., vol. 05, no. 05, pp. 74–97, 2017.
[11] A. Kaushik, “DSDM and ASD Agile Methodologies,” Int. Adv. Res. J. Sci. Eng. Technol., vol. 3, no. 9, pp. 2393–2394, 2016.

Determining symmetry groups ongeometric builds using reflection and rotation tests

Rizki Kurniawan Rangkuti[1], Raden Sri Ayu Ramadhana[2], Wahyu Azhar Ritonga[2], Muhammad Fauzi Romadhon Marpaung[2], Syabrina Rasyid Munthe[2] and Ronal Watrianthos[2*]

[1]Department of Mathematics Education, Universitas Al-Washliyah Labuhanbatu, Indonesia.
[2]Department of Informatics, Universitas Al-Washliyah Labuhanbatu, Indonesia.

* ronal.watrianthos@gmail.com

Abstract. This research is motivated by the researcher's desire to see the folding and rotational symmetry in geometric shapes whether they meet the symmetries for certain aspects and form a group. The method used in testing the symmetry of a group in geometric shapes is by testing reflection and rotation with a combination of linear and abstract algebra using an ordered matrix, where the geometric shapes are traced on each particular aspect until they meet the symmetry. From the results of the method used, it is found that the triangles to the octagon groups are symmetrical through the test of rotation and reflection of the geometric shapes. Thus, it can be concluded that group symmetry occurs up to the order of eight in geometric shapes. Based on the pattern, starting from the pentagon it is not a symmetry group, the hexagon includes the symmetry group, the heptagon is not a symmetry group, the octagon includes a symmetry group, then the nonagon is not a symmetry group and the decagon belongs to the symmetry group

1. Introduction

This research is the development of the symmetry group sub-material. It contains a symmetric group that reflects a plane figure that meets rotational and folding symmetry by involving the permutation set of an ordo matrix using binary operations (*). It meets reclusive, associative, and there is an identity matrix with each permutation matrix that has the opposite [1]. When fulfilling some of these traits, it is said that a permutation set forms a group with a binary operation. The binary operation is intended to define the multiplication of the two sets in such a way that the results obtained remain on elements of that group. A permutation set is a matrix with a certain order that can be operated with other permutation sets. The plane figures tested show that each vertex in the rotational symmetry will meets each other vertex until it return to its original point. Furthermore, in the folding symmetry vertex meets each other vertex until there is no vertex that does not occupy another vertex. The concept built on symmetry begins with the theory of reflection and rotation.Geometry is an instrument to strengthen the concepts which connect various fields in mathematics [2]. The concept formed in geometry is an elaboration of various fields of mathematics, therefore to define a new theory it has references from the matrix, algebra, and plane figure. The results obtained from the proof by reflection and rotation provide theoretical contributions to abstract algebra material which can then be used in various disciplinary applications.

However, there has been previous research [2] which states thatfor the symmetrical set of a triangular n-triangles with rotations of 120^0 and 240^0are 2^(n+1)-1sycle for n∈N while with a rotation of 360^0 is 3(2^(n+1)-1) cycle for n∈N. In addition, reflection on the S_1, S_(2,) S_3 is the 3.2^n-2 cycle for n∈N. (2) The symmetry set of n rectangles with rotations of 90^0 and 270^0 are 1/2(3^(n+1)-1) cycle, for n∈N the rotation of 180^0 is 3^(n+1)-1 cycle for n∈N and with the rotation of 360^0 is 2(3^(n+1)-1) cycle for n∈N. Whereas for reflection on the axis S_1, S_2 is 3^(n+1)-n-2 cycle for n∈Nand for reflection on the S_3, S_4 is 3^(n+1)-1 cycle for n natural number [3].

2. Methodology

2.1 Group Definition

The G-blank set is said to be a group if in G there is a binary operation stated in " * ", as according to Herstein[4]

1. For each $a, b, c \in G$ result $a * (b * c) = (a * b) * c$ (associative)
2. There is such an element $e \in G$ therefore $a * e = e * a = a$ for each $a \in G$ ($e$is the identity element in G).
3. For each $a \in G$, there is an element $a^{-1} \in G$ therefore $a * a^{-1} = a^{-1} * a = e$ (a^{-1} is the inverse of a in G).

2.2 Cayley Table

The concept of a set was first invented by a German mathematician, namely George Cantor (1918), in the late 19th century[5]. In a group, it always involves only one specific operation. Defining operations on a non-blank set is one of the terms to construct a group structure. Operation on a set can be defined by drawing a Cayley table that contains the results of the operations for each of the two elements in the set. The binary composition can be defined analytically through Cayley's table, the Cayley table is a list designed by Arthur Cayley in the 19th century[6].

2.3 Rotation and reflection

Rotation is the process of rotating a geometry that is determined by the direction and angle of rotation, while reflection is a reflection of geometry using the mirror image properties[7].

2.4 Library research method

The method in this study is the library-research method. It conducts research to obtain data and information and objects used in the problem discussion. This research relates scientific reasoning arguments to explain the results of research problems. This is in line with the opinion that literature studies are theoretical studies, references, and scientific literature relating to cultures, values, and norms that develop in the social situations studied [8].

3. Result and Discussion

Supposing that G is a non-blank set, the operation (*) on all G members is called binary operations if every a*b∈G, then a*b∈G, it can also be said that the operation (*) on G is reclusive. The reclusive characteristic here is intended that all the results of each member's binary operations remain with the members themselves[9]. A permutation is a one-to-one mapping (bijective function) from set A to the set itself [10]. If the set A={1,2,3,...,n} then the following function is a permutation.

$$1 \rightarrow f(1) = j_1$$
$$2 \rightarrow f(2) = j_2$$
$$3 \rightarrow f(3) = j_3 \quad (1)$$

∴ until

$$n \rightarrow f(n) = j_n$$

The bijective function is an injective dan surjective function[11]. Iff bijective and for$j_i \in Ai \in$ $1,2,3,\dots,n$then the permutation is given with the following two line notations:

$$\begin{bmatrix} 1 & 2 & 3 \dots & n \\ j_1 & j_2 & j_3 \cdots & j_n \end{bmatrix} \quad (2)$$

If given A={1,2,3}. The permutation can be indicated by:

$$\begin{aligned} f:\ 1 &\to f(1) = 1g:\ 1 \to g(1) = 2 \\ 2 &\to f(2) = 22 \to g(2) = 3 \\ 3 &\to f(3) = 33 \to g(3) = 1 \end{aligned} \quad (3)$$

Wrtitten by $f = \begin{bmatrix} 1 & 2 & 3 \\ 1 & 2 & 3 \end{bmatrix}$Written by $g = \begin{bmatrix} 1 & 2 & 3 \\ 2 & 3 & 1 \end{bmatrix}$

$$\begin{aligned} h:\ 1 &\to h(1) = 3 \qquad i:\ 1 \to i(1) = 1 \\ 2 &\to h(2) = 12 \to i(2) = 3 \\ 3 &\to h(3) = 23 \to i(3) = 2 \end{aligned} \quad (4)$$

Wrtitten by $h = \begin{bmatrix} 1 & 2 & 3 \\ 3 & 1 & 2 \end{bmatrix}$Wrtitten by$i = \begin{bmatrix} 1 & 2 & 3 \\ 1 & 3 & 2 \end{bmatrix}$

$$\begin{aligned} j:\ 1 &\to j(1) = 3 \qquad k:\ 1 \to k(1) = 2 \\ 2 &\to j(2) = 22 \to k(2) = 1 \\ 3 &\to j(3) = 13 \to k(3) = 3 \end{aligned} \quad (5)$$

Wrtitten by $j = \begin{bmatrix} 1 & 2 & 3 \\ 3 & 2 & 1 \end{bmatrix}$Wrtitten by$k = \begin{bmatrix} 1 & 2 & 3 \\ 2 & 1 & 3 \end{bmatrix}$

So there are six permutation sets of A={1,2,3}, or it can be determined P_3^3which states many sets of permutations.The permutation set can also be written in the form of a cycle.

$$\begin{aligned} f = \begin{bmatrix} 1 & 2 & 3 \\ 1 & 2 & 3 \end{bmatrix} & g = \begin{bmatrix} 1 & 2 & 3 \\ 2 & 3 & 1 \end{bmatrix} \\ 1 \to f(1) = 1&1 \to g(1) = 2 \\ 2 \to f(2) = 2&2 \to g(2) = 3 \\ 3 \to f(3) = 3&3 \to g(3) = 1 \end{aligned} \quad (6)$$

Suppose the function$f: A \to B$ determined by rule $f(a) = b$, while the function$f: B \to C$ determined by rule$f(b) = c$. The composition functions$g$ and f written by gof function defined by rule$(gof)(a) = g(f(a))$ [12].The permutation of fog and gof can be given by:

$$fog = \begin{bmatrix} 1 & 2 & 3 \\ 1 & 2 & 3 \end{bmatrix}\begin{bmatrix} 1 & 2 & 3 \\ 2 & 3 & 1 \end{bmatrix} \qquad gof = \begin{bmatrix} 1 & 2 & 3 \\ 2 & 3 & 1 \end{bmatrix}\begin{bmatrix} 1 & 2 & 3 \\ 1 & 2 & 3 \end{bmatrix}$$

$$\begin{aligned} 1 \to g(1) = 2 \to 2 &\qquad 1 \to f(1) = 1 \to 2 \\ 2 \to g(2) = 3 \to 3 &\qquad 2 \to f(2) = 2 \to 3 \\ 3 \to g(3) = 1 \to 1 &\qquad 3 \to f(3) = 3 \to 1 \end{aligned} \quad (7)$$

It is known that A={1,2,3} there are six permutations. The permutation set A is f, g, h, i, j, k whose inverse that:

$$\begin{aligned} f = \begin{bmatrix} 1 & 2 & 3 \\ 1 & 2 & 3 \end{bmatrix} \text{ invers } f = \begin{bmatrix} 1 & 2 & 3 \\ 1 & 2 & 3 \end{bmatrix} &\qquad i = \begin{bmatrix} 1 & 2 & 3 \\ 1 & 3 & 2 \end{bmatrix} \text{ invers } i = \begin{bmatrix} 1 & 2 & 3 \\ 1 & 3 & 2 \end{bmatrix} \\ g = \begin{bmatrix} 1 & 2 & 3 \\ 2 & 3 & 1 \end{bmatrix} \text{ invers } h = \begin{bmatrix} 1 & 2 & 3 \\ 3 & 1 & 2 \end{bmatrix} &\qquad j = \begin{bmatrix} 1 & 2 & 3 \\ 3 & 2 & 1 \end{bmatrix} \text{ invers } j = \begin{bmatrix} 1 & 2 & 3 \\ 3 & 2 & 1 \end{bmatrix} \\ h = \begin{bmatrix} 1 & 2 & 3 \\ 3 & 1 & 2 \end{bmatrix} \text{ invers } g = \begin{bmatrix} 1 & 2 & 3 \\ 2 & 3 & 1 \end{bmatrix} &\qquad k = \begin{bmatrix} 1 & 2 & 3 \\ 2 & 1 & 3 \end{bmatrix} \text{ invers } k = \begin{bmatrix} 1 & 2 & 3 \\ 2 & 1 & 3 \end{bmatrix} \end{aligned} \quad (8)$$

Members of the set A={1,2,3} can be supposed as three dots on a triangle, we will test whether to form a group with multiplication operations (*) through reflection (frame folds) and rotation (permutations). We know that members of set A are permutations so that we can write them down by:

$$\begin{aligned} f = \begin{bmatrix} 1 & 2 & 3 \\ 1 & 2 & 3 \end{bmatrix} = (1)(2)(3) &\qquad i = \begin{bmatrix} 1 & 2 & 3 \\ 1 & 3 & 2 \end{bmatrix} = (1)\ (2\ 3) \\ g = \begin{bmatrix} 1 & 2 & 3 \\ 2 & 3 & 1 \end{bmatrix} = (1\ 2\ 3) &\qquad j = \begin{bmatrix} 1 & 2 & 3 \\ 3 & 2 & 1 \end{bmatrix} = (1\ 3)\ (2) \\ h = \begin{bmatrix} 1 & 2 & 3 \\ 3 & 1 & 2 \end{bmatrix} = (1\ 3\ 2) &\qquad k = \begin{bmatrix} 1 & 2 & 3 \\ 2 & 1 & 3 \end{bmatrix} = (1\ 2)(3) \end{aligned} \quad (9)$$

3.1 Triangle

Triangle shown in figure 1 below that has points A,B,C can be inserted frame in 6 ways, which case it can be said that there are 6 transformations so that the triangle is the same side through reflection there are 3 ways and rotation there are 3 ways. In this case,the vertices of triangle can be written by A,B,C which is in the member of our permutation set with 1,2,3.

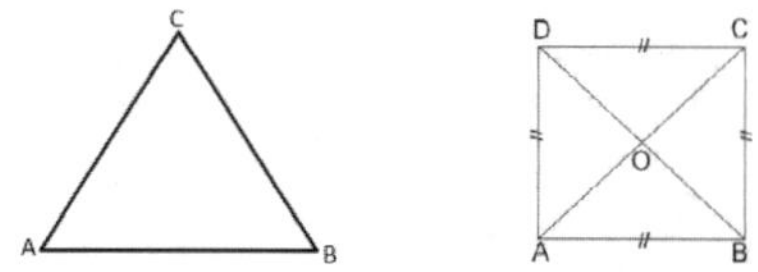

(a) (b)

Figure 1. (a) Equilateral Triangel, (b) Equilateral Rectangel

The three reflections in Figure 1 are the reflections on the AX, BY, dan CZ. Reflection AX, point A is shrimp but B and C shipment each other, the permutations that occur are:

$$i = \begin{bmatrix} 1 & 2 & 3 \\ 1 & 3 & 2 \end{bmatrix} = (2\ 3) \tag{10}$$

Reflection BY, B is shimpit but A and C shimpit each other, the permutations that occur are:

$$j = \begin{bmatrix} 1 & 2 & 3 \\ 3 & 2 & 1 \end{bmatrix} = (1\ 3) \tag{11}$$

Reflection CY, C shimpit but A and B shimpit each other, then the permutations that occurare:

$$k = \begin{bmatrix} 1 & 2 & 3 \\ 2 & 1 & 3 \end{bmatrix} = (1\ 3) \tag{12}$$

The three rotations (permutations) are the rotation of the field with the center of the O and the direction of the rotation in the opposite direction of the clock rotation. Rotation with an angle360^0, A rotates back at its, B and C, then the permutations that occur are:

$$f = \begin{bmatrix} 1 & 2 & 3 \\ 1 & 2 & 3 \end{bmatrix} \tag{13}$$

Rotation with an angle120^0, A rotates one-third of the angle of one full round, B and C, sothe permutations that occur are:

$$g = \begin{bmatrix} 1 & 2 & 3 \\ 2 & 3 & 1 \end{bmatrix} = (1\ 2\ 3) \tag{14}$$

Rotation with an angle240^0, A rotates two-thirds of the angle of one full round, B and C, so the permutations that occur are:

$$h = \begin{bmatrix} 1 & 2 & 3 \\ 3 & 1 & 2 \end{bmatrix} = (1\ 3\ 2) \tag{15}$$

Multiplication A*A can be shown in cayley table below:

Table 1. Multipication Cayley Table ofPermutation Seton Triangle

(A,*)	***f***	***g***	***h***	***i***	***j***	***k***
f	*f*	*g*	*h*	*i*	*j*	*k*
g	*g*	*h*	*f*	*k*	*i*	*j*
h	*h*	*f*	*g*	*j*	*k*	*i*
i	*i*	*j*	*k*	*f*	*g*	*h*
j	*j*	*k*	*i*	*h*	*f*	*g*
k	*k*	*i*	*j*	*g*	*h*	*f*

A group is empty space that fulfills reclusive, associative, has an element of identity and inverse. From the table above can be proven A*A which is empty space forming a group[13]:

- reclusive, indicated from $A*A$ included A
- associative, indicated by taking section A where the result of the operation shows the right field the same as the left

$$f * (g * h) = (f * g) * hi * (j * k) = (i * j) * k$$
$$f * f = g * hi * g = g * k \tag{16}$$
$$f = fj = j$$

- the element of identity $\forall g \in A\ \exists f \in A \ni g * f = g$ (17)
- invers $\exists g^{-1} \in A \forall g \in A \ni g^{-1} * g = f$ (18)

$$\exists g^{-1} \in A \forall g \in A \ni \quad h * g = f$$

The fulfillment of A*A reclusive, associative, an element of identity and inverse, then (A*) is a group. Because the ABC triangle meets rotational symmetry and reflection, the triangle is a symmetry group.

3.2 Rectangle

Rectangle with the same side (equilateral rectangle) of A, B, C, D as seen in figure 2 can be inserted the frame in 8 ways, in this case, it can be said that there are 8 transformations so that the rectangle through reflection there are 4 ways and the rotation there are 4 ways.Using Cayley table for *G*G* multiplication:

Tabel 2. Multipication Cayley Table of Permutation Seton Rectangel

(G,*)	***I***	***R***	R^2	R^3	***A***	***B***	***C***	***D***
I	I	R	R^2	R^3	A	B	C	D
R	R	R^2	R^3	I	C	D	B	A
R^2	R^2	R^3	I	R	B	A	D	C
R^3	R^3	I	R	R^2	D	C	A	B
A	A	D	B	C	I	R^2	R^3	R
B	B	C	A	D	R^2	I	R	R^3
C	C	A	D	B	R	R^3	I	R^2
D	D	B	C	A	R^3	R	R^2	I

From the table above can be proven whether *G*G* forms a group because

- reclusive, indicated from*G*G* included*G*
- associative, indicated by taking section G where the result of the operation shows the right field the same as the left

$$R * (R^3 * D) = (R * R^3) * DR * (A * C) = (R * A) * C$$
$$R * B = I * DR * R^3 = C * C \quad (19)$$
$$D = DI = I$$

- the element of identity $\forall R^2 \in G \ \exists I \in G \ni R^2 * I = R^2$ (20)
- invers $\exists C^{-1} \in G \forall C \in G \ni C^{-1} * C = I$ (21)

$$\exists C^{-1} \in G \forall C \in G \ni \quad C * C = I$$

The fulfillment of G*G reclusive, associative, element of identity, andinverse, then (G*) is a group. Therefore the ABCD rectangle is a symmetry group indicated by rotation and reflection.

3.3 Pentagon and Hexagon

By following the pattern of symmetry through reflection and rotation tests, the same hexagon side A, B, C, D, E cannot be inserted in the frame, as some points cannot be reflected the other point, meaning it is not filled with folding symmetry.

Based on figure 2 when point A reflected C, but E and D do not attach each other. The evidence shows that the hexagon is not a symmetry group. The hexagon sides of A, B, C, D, E, F as seen in figure 2 can be inserted in the frame, meaning that symmetry is filled. As seen in figure 4. If point A is reflected B, C with F, and D enclosed with E. If A reflected C, then D enclosed with F then B and E are attached.

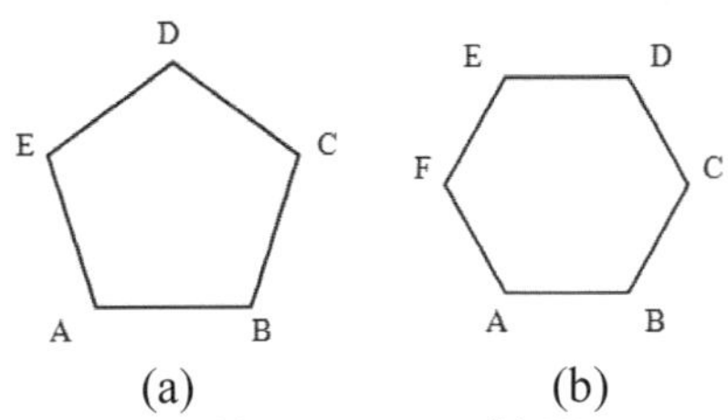

Figure 2. (a) Pentagon, (b) Hexagon

3.4 Heptagon and Octagon

By followingthe symmetry through reflection and rotation tests, then the octagon A,B,C,D,E,F,G cannot be inserted in the frame, since some points cannot be reflected another point, meaning that the symmetry is not filled.

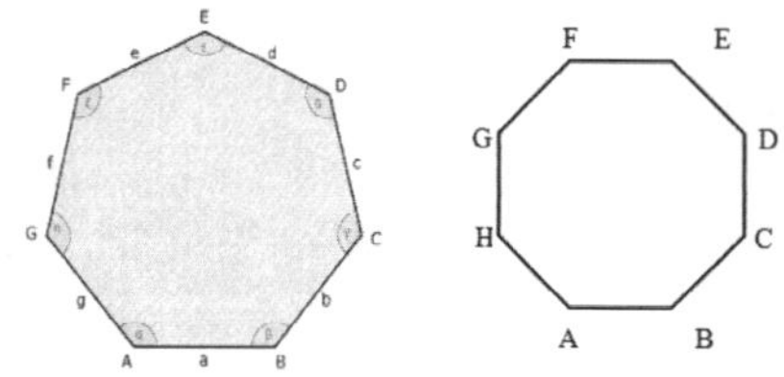

(a) (b)

Figure 3. (a) Heptagon, (b) Octagon

From Figure 5 it can be seen when point A is reflected to point C, point E, and point D are not attached to each other. The evidence can already show that the octagon is not a symmetry group. In figure 6, the octagon A, B, C, D, E, F, G, H can be inserted into the frame, meaning that the folded symmetry is filled, and an octagonal point equal to the side A, B, C, D, E, F, G, H can be rotated and occupy the other points on the octagon on the same side. It means the rotation symmetry is filled. Thus the octagon A,B,C,D,E,F,G,H is defined as a symmetry group.

4. Conclusion

Based on the pattern, starting from the pentagon it is not a symmetry group, the hexagon includes the symmetry group, the heptagon is not a symmetry group, the octagon includes a symmetry group, then the nonagon is not a symmetry group and the decagon belongs to the symmetry group. Finally, it can be concluded that the n-facets of a plane figure with n>3 and n odd numbers are not symmetry groups and n-facets in a plane figure with n>3 and n even numbers are symmetry groups.

References

[1] Syahrir, "Pengembangan Pola Berpikir Siswa Kelas Xi Tentang Pengoprasian Dasar Pada Matriks Identitas melaui metode pembelajaran Tanya jawab," *J. Ilmu Sos. dan Pendidik.*, vol. 2, no. 1, 2018.

[2] K. Safrina, M. Ikhsan, and A. Ahmad, "Peningkatan Kemampuan Pemecahan Masalah Geometri melalui Pembelajaran Kooperatif Berbasis Teori Van Hiele," *J. Didakt. Mat.*, vol. 1, no. 1, 2014.

[3] J. Wang, X. G. Wen, and E. Witten, "Symmetric Gapped Interfaces of SPT and SET States: Systematic Constructions," *Phys. Rev. X*, 2018, doi: 10.1103/PhysRevX.8.031048.

[4] N. R. Dewi, N. Eliyati, and O. H. Marbun, "Kajian Struktur Aljabar Grup pada Himpunan Matriks yang Invertibel," *J. Penelit. Sains*, vol. 14, no. 1, 2011.

[5] K. W. Johnson and J. D. H. Smith, "On the category of weak Cayley table morphisms between groups," *Sel. Math.*, vol. 13, no. 1, pp. 57–67, Jun. 2007, doi: 10.1007/s00029-007-0032-x.

[6] G. Cooperman and L. Finkelstein, "New methods for using Cayley graphs in interconnection networks," *Discret. Appl. Math.*, 1992, doi: 10.1016/0166-218X(92)90127-V.

[7] D. Setyawan, "Eksplorasi Proses Konstruksi Pengetahuan Materi Bangun Ruang Siswa Dengan Gaya Berpikir Acak Dan Kemampuan Keruangan Level Rotasi Mental," *Ecosystem*, vol. 17, no. 1, 2017.

[8] Sugiyono, *Memahami Penelitian Kualitatif.* Bandung: Alfabeta, 2012.

[9] R. S. Waskitoningtyas, *MENGENAL MATEMATIKA DISKRIT*. Cirebon: Nusa Litera Inspirasi, 2019.

[10] R. T. Damayanti, "Automorfisme Graf Bintang Dan Graf Lintasan," *Xcauchy*, vol. 2, no. 1, 2011.

[11] 020 Umy Zahroh, "Kecenderungan Gaya Belajar Mahasiswa dalam Menyelesaikan Masalah Fungsi Bijektif," *J. Kebijak. dan Pengemb. Pendidik.*, 2014.

[12] Siswanto & Supraptinah, *Matematika Inovatif 2; Konsep dan Aplikasinya*. Pusat Perbukuan Depdiknas Tahun 2009, 2009.

[13] A. J. Noor and N. Hijriati, "GRUP RING," *J. Mat. Murni dan Terap.*, vol. 4, no. 1, 2010.

Online learning sentiment analysis during the covid-19 Indonesia pandemic using twitter data

Syafrida Hafni Sahir[1], Raden Sri Ayu Ramadhana[2], Muhammad Fauzi Romadhon Marpaung[2], Shabrina Rasyid Munthe[2] and Ronal Watrianthos[2*]

[1]Department of Management, Universitas Medan Area, Indonesia
[2]Department of Informatics, Universitas Al Washliyah Labuhanbatu, Indonesia

* ronal.watrianthos@gmail.com

Abstract. In the last week of September 2020, Covid-19 in Indonesia has infected more than 252,000 people. The spread of the virus through physical contact forces all countries to use social distance and physical distance to reduce interactions. On the educational side, the Covid-19 pandemic has had an impact on the prevention of face-to-face teaching and learning activities. All educational institutions must stop the learning process and replace it with online learning. Nadiem Makarim, Minister of Education and Culture of the Republic of Indonesia, then affirmed this online learning by issuing a policy on online learning in case of an emergency spread of the Covid-19 virus. However, this online learning has provoked debate in the community over the preparation of the technology. The goal of this study was to analyze public opinion for online learning during the Covid-19 pandemic in October 2020. The research carried out document-based text mining and feelings on Twitter data analyzed using the Naïve Bayes method. Analysis in that period found 25 percent positive sentiment, 74 percent negative sentiment, and 1 percent neutral sentiment.Several tweets showed that the words 'stress' and 'covid' were the most commonly spoken in the month.

1. Introduction

The World Health Organization (WHO) declared the COVID-19 virus a global pandemic on March 11, 2020 [1]. WHO reports 35 million people infected with Covid-19 and 1 million more people died in early October 2020. Meanwhile, Indonesia recorded 307 thousand people confirmed positive with the death toll has reached 11,250 people [2]. The rapid spread of the virus through physical contact forces all countries to apply social distancing and physical distancing to reduce interactions. The main factor for transmission through droplets released during talking, coughing, or sneezing makes social distancing the most widely adopted strategy when there is no vaccine [3].

Indonesia's President Jokowi has issued a Large-Scale Social Restriction (PSBB) policy [4]. Social restrictions affect public spaces, including offices, schools, or campuses. In education, this strategy prohibits face-to-face learning and forces all educational institutions to stop the learning process and replace it with online learning [5]. In education, difficulties emerge, such as the unpreparedness of technologies, the media used, or the psychological aspects of students. In the meantime, the future will be unequal education between populations and regions in Indonesia in the long term [6].

The plan to tackle the pandemic of social constraints pushes all educational institutions to embrace online learning [5]. The Ministry of Education and Culture of the Republic of Indonesia Nadiem Makarim underscored the policy of online learning during the pandemic by issuing the Learning From

Home (BDR) policy [7]. This policy requires the use of the Internet network through the intermediary of smartphones, gadgets, computers, and applications as a substitute for face-to-face use.

Online learning, which is called a strategy, is then problematic because it has to be adapted. Infrastructure gaps, the efficiency of connections, devices used, and internet prices are the biggest hurdles. The abrupt transition from face-to-face learning to large-scale online learning has given rise to a diverse variety of responses in public. This study aims to analyze online education public sentiment during the October 2020 pandemic of Indonesia COVID-19. This research uses the keywords 'online learning, lecture, learning,' and the hashtag '#BelajarDariRumah' which are filtered with the keyword 'online' and 'home' on Indonesian tweets which are scrambled from Twitter and processed using the Naïve Bayes algorithm to produce sentiment analysis on online learning acceptance.

2. Methodology

2.1 Sentiment Analysis

Analysis of sentiment is a method to recognize opinion and classify text in a document/phrase to identify groups as positive, negative, or neutral [8]. Today, academics research is commonly using emotion analysis as a branch of computer science research. Social networks, such as Twitter, are widely used in sentiment analysis to classify public views [9].

It is also possible to equate sentiment analysis with opinion mining since it relies on viewpoints that are favorable or negative [10]. Data mining is applied to analyze, process, and retrieve textual data within an object in sentiment analysis, such as utilities, goods, persons, phenomena, or other subjects. The research process can involve text reviews, forums, tweets, or blogs, with pre-processing data covering tokenization, stopword, deletion, stemming, identification of emotions, and classification of sentiments [11].

2.2 Text Mining

Text mining is known to produce information from a set of documents. Text Mining can generate information by processing, grouping, and analyzing large volumes of unstructured data [12]. Data mining is to extract valuable information from a variety of documents with unstructured text data sources [13]. The process of collecting data in text mining can produce a sensing analysis that identifies an emotionally positive or negative statement [14].

2.3 Naïve Bayes Method

The Naïve Bayes method is a text mining classification method used for the study of emotion. In terms of data consistency and measurement classification, this method is technically reasonable [15]. In the Twitter classification method, Naïve Bayes uses a variety of approaches, including Unigram Naïve Bayes, Multinomial Naïve Bayes, and Maximum Entropy Classification [16]. The main feature of the Naïve Bayes classification is the firm hypothesis of each situation or event [17]. Calculation of the probability groups of Naïve Bayes uses the Bayes algorithm method using the following equation:

$$P(c|z) = \frac{p(z|c)\ X\ p(c)}{p(z)} \quad (1)$$

Equation (1) shows that Y is a specific class, X is data in the unknown classed, while P (Y | X) is the probability of the hypothesis based on conditions, while P (Y) and P (X | Y) is the previous probabilities of the class based on hypothetical, while P (X) is the probability Y [18]. In the Naïve Bayes classification, equation 1 is further developed into equation (2) as follows:

$$P(Y|X1, X2, \ldots Xn) = \frac{P(X1, X2, \ldots Xn|Y)P(Y)}{P(X1, X2, \ldots Xn)} = \frac{P(X1|Y)P(X2|Y)\ldots P(Xn|Y)P(Y)}{P(X1, X2 \ldots Xn)} \quad (2)$$

Where $P(Y|X1, X2, \ldots Xn)$ is the result of the calculation of all posterior probabilities at the value of X for all Variable y, the Naïve Bayes classification will make predictions dependent on the maximal likelihood of the posterior Laplace probability shown in equation (3) below, where c is the number of Y values [21]:

$$P(Xi|Y) = \frac{Nic+1}{Nc+c} \quad (3)$$

Figure 1 shows an opinion mining method using the Naïve Bayes process on Twitter. The process of crawling twitter data in this analysis is to provide keywords at a certain period. Labeling is applied as data is collected to determine emotions. The next step is the pre-processing and systematic transformation of data collection. At this point, a cleaning process is performed out to minimize noise and delete stopwords to eliminate useless words such as 'I,' 'and' or ' [5]. The tokenization method is used to define words and split sentences into space-based phrases and punctuation marks. The last step of pre-processing is to convert the affixes to simple phrases. The third stage of opinion mining is a feature extract intended to ease the classification of Naïve Bayes. This stage produces a model which is used to show the consistency of the effects of the classification [19].

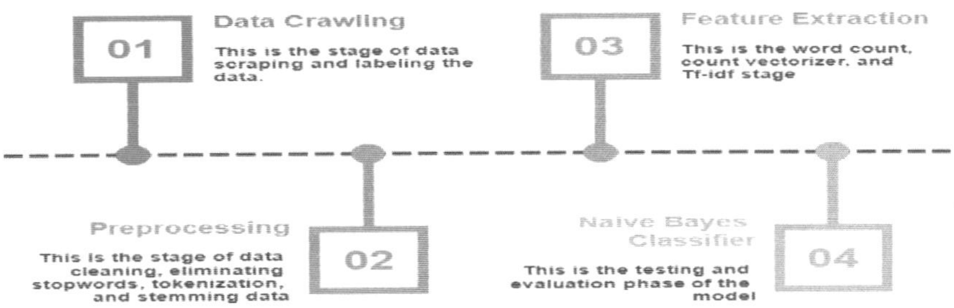

Figure 1. The Naïve Bayes Process for Classifying

The results of the crawling tweets are stored in a CSV file and split into two datasets, namely training data and test data. Positive and negative tweets are categorized on a label list. The Naïve Bayes method will be used at the stages of sentiment classification and interpretation of the result of the sentiment analysis.

3. Result and Discussion

We use the terms of online studying, lecture, research, and hashtag # BelajarDariRumah as keywords for Indonesian tweets. The data crawling method using the public Twitter stream of Drone Emprit Academic was able to capture 159,045 tweets linked to the chosen keywords shown in Table 1 [20].

Table 1. Crawling Data

@aliefriendshew	Sekolah online : tambah bego Sekolah offline : meresahkan Yaallah capek bgt akutu
@nesshlyn	Buat yg kelas 9 sama 12 ngerasa ngga? Udah nyaman sekolah online, kalau offline pun udah males duluan. Tapi kepikiran pelajaran sama sekali ngga ngerti takut buat kedepanya plonga-plongo
@xxzaww	saya lebih menyukai sekolah offline dari pada online,terimakasih
@schfess	Sch! Semangat buat kamu yg tugasnya masih numpuk walopun libur!! kita semua disini sama-sama berjuang sekolah online jadi jgn ngerasa sendirian/putus asa ya!! https://t.co/lBf2170OFM
@dulce_maria_lrs	Gue pribadi cukup miris dengan guru yang kasih tugas berlebihan untuk murid, dengan alasan sekolah daring. Bayangin aja,sekolah online by zoom,gru jelasin, dan dikasih tugas setumpuk lagi, itu baru 1 mapel, sedang sehari 3-4 mapel, lu bayangin itu otak kerjanya kayak apa+

The process of tokenization and data conversion occurs during preprocessing to remove noise and clarify features. The data cleansing stage removes unnecessary attributes such as URL, mention, username, RT, hashtag, or punctuation and class assignments. The preprocessing results were grouped to get the most word frequency and visualized into positive and negative word cloud shown in Figure 2.

Figure 2. Wordclouds Positive and Negative

The sentiment analysis model is built based on training data that has been prepared and tested to determine sentiment towards online learning. The text analyzed in this study is the text in Indonesian tweets and uses Indonesian vocabulary [21]. The results of the sentiment test data analysis on the training data could be in Table 2 for five sample data.

Table 2. Test Data Analytical Results

Sentiment	Confident (Positive)	Confident (Neutral)	Confident (Negative)	Text (in Bahasa)
Positive	0.9856	0.0071	0.0073	Semangat buat yang lagi sekolah online kuliah online, ngerjain tugas, magang, kerja dari rumah maupun harus keluar, dan kegiatan apapun. Tetap hati2 dan jaga kesehatan ya!
Negative	0.0072	0.0070	0.9860	kuliah online bikin weekend jadi ga punya harga diri lagi like, sekarang mau hari kerja atau akhir pekan ga ada bedanya, full meeting, full nugas, full deadline semua stay hydrated all
Negative	0.0076	0.0074	0.9877	Kendala kuliah online gue tiap bimbingan gue paling sebel kalau pas lagi asistensi terus tiba2 emak gue buka pintu triak nyuruh cuci piring pdhl lagi vidio call sama dospem
Negative	0.0071	0.0073	0.9871	lebih ke capek, rutinitas kuliah online kerasa kalau gitu gitu aja soalnya. kuliah-nugas-kuliah-nugas dan itu harus ngadep laptop terus. lama lama jenuh terus capek aja rasanya. capek mata sama capek badan juga
Positive	0.0989	0.0070	0.0075	jaman sekarang kalo mau belajar mah udah enak ya, bisa belajar online diajarin tutor yang ramah nan ceria, jaman w dulu belajar di rumah diajarin emak, digembleng ampe nangis2

Just 25 percent had a positive sentiment based on the findings of the study tested to get a sense of online learning on Twitter in the October period. At 75 percent, negative opinion dominates, and the remaining 1 percent is neutral.

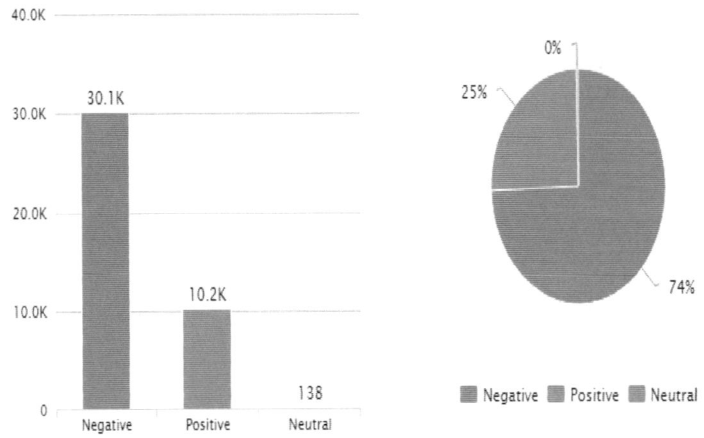

Figure 3. Measurement of sentiment with Naïve Bayes.

Figure 3 shows the sentiment measurement results with more than 30 thousand tweets being negative, more than 10 thousand tweets giving positive results, and the remaining 138 tweets were considered neutral during the month. Based on these results, online learning has not been maximally applied in Indonesia during the pandemic, as seen from the high public disappointment.

4. Conclusion

The Naïve Bayes algorithm for sentiment classification applies to this analysis. Sentiment analysis in this research uses Twitter data with the keywords 'online learning' 'school' 'studying' and the hashtag #BelajarDariRumah filtered with the keywords 'online' and 'home' on Indonesian tweets in October 2020. Analysis in that period found 25 percent positive sentiment, 74 percent negative sentiment, and 1 percent neutral sentiment. People's frustration with online learning produced negative perceptions. Several tweets revealed discontent with the words 'stress' and 'covid' having the highest communication level throughout the month. In future experiments, the use of various algorithms is likely to provide more reliable performance.

References

[1] D. Cucinotta and M. Vanelli, "WHO declares COVID-19 a pandemic," *Acta Biomed.*, vol. 91, no. 1, pp. 157–160, 2020, doi: 10.23750/abm.v91i1.9397.

[2] WHO, "WHO Coronavirus Disease (COVID-19) Dashboard," *WHO Health Emergency Dashboard*, 2020. https://covid19.who.int/ (accessed Oct. 07, 2020).

[3] D. M. Dave, A. I. Friedson, K. Matsuzawa, J. J. Sabia, and S. Safford, "Black lives matter protests, social distancing, and COVID-19," *NBER Work. Pap. Ser.*, 2020.

[4] A. Amindoni, "Virus corona: Presiden Jokowi pilih 'pembatasan sosial dalam skala besar', warga mulai sortir pendatang," *BBC News Indonesia*, 2020. https://www.bbc.com/indonesia/indonesia-52059236 (accessed Oct. 07, 2020).

[5] R. Watrianthos, "Analisis Pembelajaran Daring di Era Pandemic Covid-19," in *Merdeka Kreatif di Era Pandemi Covid-19: Suatu Pengantar*, Medan: Green Press, 2020, p. 55.

[6] R. H. Syah, "Dampak Covid-19 pada Pendidikan di Indonesia: Sekolah, Keterampilan, dan Proses Pembelajaran," *SALAM J. Sos. dan Budaya Syar-i*, vol. 7, no. 5, Apr. 2020, doi: 10.15408/sjsbs.v7i5.15314.

[7] Kumparan, "4 Kebijakan Nadiem Makarim soal Proses Belajar dari Rumah Selama Pandemi Corona," *kumparan.com*, 2020. https://kumparan.com/kumparanmom/4-kebijakan-nadiem-makarim-soal-proses-belajar-dari-rumah-selama-pandemi-corona-1t5naOVW9MB/full (accessed Oct. 07, 2020).

[8] D. A. Ramadhan and E. B. Setiawan, "Analisis Sentimen Program Acara Di Sctv Pada Twitter

Menggunakan Metode Naive Bayes Dan Support Vector Machine," in *e-Proceeding of Engineering*, 2019, pp. 9376–9743.

[9] R. Ferryawan, Kusrini, and F. W. Wibowo, "Analisis Sentimen Wisata Jawa Tengah Menggunakan Naïve Bayes," *J. Inf. Politek. Indonusa Surakarta*, vol. 5, no. 3, pp. 55–60, 2019.

[10] C. B. Saputra, A. Muzakir, and D. Udariansyah, "Analisis Sentimen Masyarakat Terhadap #2019gantipresiden Berdasarkan Opini Dari Twitter Menggunakan Metode Naive Bayes Classifier," in *Bina Darma Conference on Computer Science*, 2019, pp. 403--413.

[11] S. B. Bhonde and J.R. Prasad, "Sentiment Analysis-Methods, Applications and Challenges," *nternational J. Electron. Commun. Comput. Eng.*, vol. 6, no. 6, pp. 634–640, 2015.

[12] A. Sholihin, Haviluddin, N. Puspitasari, M. Wati, and Islamiyah, "Analisis Penyakit Difteri Berbasis Twitter Menggunakan Algoritma Naïve Bayes," *SAKTI – Sains, Apl. Komputasi dan Teknol. Inf.*, vol. 1, no. 1, pp. 7–15, 2019.

[13] G. E. I. Kambey, Rizal Sengkey, and Agustinus Jacobus, "Penerapan Clustering pada Aplikasi Pendeteksi Kemiripan Dokumen Teks Bahasa Indonesia," *J. Tek. Inform.*, vol. 15, no. 2, pp. 75–82, 2020.

[14] X. Zhou, X. Tao, and Z. Yang, "Sentiment Analysis on Tweets for Social Events," in *IEEE 17th International Conference on Computer Supported Coorporative Work in Design*, 2013, pp. 557–562.

[15] J. M and V. H, "Opinion Mining For Sentiment Data Classification," *Int. J. Res. Inf. Technol.*, vol. 3, no. 1, pp. 1–13, 2014.

[16] R. P and M. M, "Sentiment Analysis of User Generated Twitter Updates using Various Classification," 2009.

[17] N. Rochmawati and S. C. Wibawa, "Opinion Analysis on Rohingya using Twitter Data," *IOP Conf. Ser. Mater. Sci. Eng.*, vol. 336, no. 1, 2018, doi: 10.1088/1757-899X/336/1/012013.

[18] Syahriani, A. A. Yana, and T. Santoso, "Sentiment analysis of facebook comments on indonesian presidential candidates using the naïve bayes method," in *Journal of Physics: Conference Series*, Nov. 2020, vol. 1641, p. 012012, doi: 10.1088/1742-6596/1641/1/012012.

[19] R. Watrianthos, S. Suryadi, D. Irmayani, M. Nasution, and E. F. S. Simanjorang, "Sentiment Analysis Of Traveloka App Using Naïve Bayes Classifier Method," *Int. J. Sci. Technol. Res.*, vol. 8, no. 07, pp. 786–788, 2019.

[20] I. Fahmi, "Drone Emprit Academic: Software for social media monitoring and analytics," *dea*, 2020. academic.droneemprit.id (accessed Oct. 31, 2020).

[21] D. R. Lazuardi, T. A. Munandar, H. Harsiti, Z. Mutaqin, and R. N. Hays, "Sentiment analysis of public opinions on the welfare of honorary educators using Naive Bayes," *IOP Conf. Ser. Mater. Sci. Eng.*, vol. 830, p. 032018, May 2020, doi: 10.1088/1757-899X/830/3/032018.

Analysis of The Heavy Metal Content of Pb, Cu and Hg in Leachate at Final Waste Disposal Batu Bola Padangsidimpuan City

Fatma Suryani Harahap[1*], Adi Syahputra[1], Nurmaini Ginting[2] , Nasir Sah[1], and Muharram Fajri[3]

[1]Chemistry Education Study Program, Universitas Muhammadiyah Tapanuli Selatan, Indonesia.
[2]Biology Education Study Program, Universitas Muhammadiyah Tapanuli Selatan, Indonesia.
[3]Animal Husbandry Study Program, Universitas Muhammadiyah Tapanuli Selatan, Indonesia.

* fatma.suryani@um-tapsel.ac.id

Abstract. Safe and proper disposal of waste is an important component in integrated waste management. The method of accumulating garbage in landfills (TPA) can result in a decrease in environmental quality due to air pollution due to solid waste burning, groundwater pollution due to leaching, and an increasing number of disease vectors that can endanger the health of waste collectors and residents. around the landfill. The purpose of this study was to determine the metal content of Lead (Pb), Mercury (Hg) and Copper (Cu) in the leachate of TPA BatuBola.This research method uses descriptive exploratory method by conducting a survey first. Atomic Absorption Spectrophotometer is a tool used in this study.Data analysis was carried out by comparing the test result data with the quality standards by Permen LHK P.59 / 2016 and Permen LHK No.5 / 2014. The data shows that the concentration of Hg metal is <0.0004 mg / L. Meanwhile, the threshold set by the government is 0.005 mg / L. The concentration of lead metal (Pb) also shows a value that is below the quality standard threshold, namely <0.003 mg / L, while the set limit is 0.1 mg / L. As well, the measurement of copper metal (Cu) is still below the quality standard threshold, namely <0.006 mg / L, while the limit set by the government is 2.0 mg / L. However, caution must be exercised because if it is accumulated in large quantities it can still be dangerous.

1. Introduction

Leachate (leachate) is a liquid that seeps through a pile of waste carrying dissolved or suspended material, especially the result of the decomposition of waste matter[1]. Leachate is a type of pollutant that has a high potential to pollute the environment around the landfill because it contains very toxic materials[2]. Leachate that seeps into the soil will cause direct groundwater pollution[3]. The composition of leachate is influenced by several factors such as the type of waste deposited, the amount of rainfall in the landfill area and the specific conditions of the dumpsite. [4]. Leachate generally contains high levels of organic and inorganic compounds heavy metals[5]. Leachate from the landfill has the potential to contain several heavy metals, including cadmium (Cd), lead (Pb), copper (Cu), chromium (Cr), mercury (Hg) and zinc (Zn) in landfill (Final Disposal Place).TPA is a place for storing various kinds of waste so that leachate contains various types of pollutants that have the potential to disrupt the environment and human health. Leachate can seep into the ground, or flow on the ground and empties into river water flow. Each landfill has different leachate characteristics depending on the processes that occur in the landfill, which include physical, chemical and biological processes.[6]who conducted a study on the characteristics of leachate in landfills in Riyadh City, Saudi Arabia found that leachate in these locations contained Chemical Oxygen Demand (COD), Biology Oxygen Demand (BOD), Electrical Conductivity (EC), Total Suspended Solid (TSS) with a high concentration but has a low pH value. In addition, leachate in the Riyadh City TPA contains Fe, Mn, Mo, Ni, Cr, Zn and Cu. Zub[7], found that the test results of leachate samples at TPA Maros South Sulawesi had high COD, BOD and TSS values.[8]conducted a study to determine the characteristics of leachate in TPA Air Hitam in Puchong Selangor and found that leachate had very low COD and relatively high heavy metal concentrations for Fe, Zn, Pb and Ni while Mn, Cd, Cu and Cr were found to be relatively low.

The disposal system applied to the Batu Bola landfill in Padangsiidmpuan City is an open dumping system, where garbage is piled up without any geotextile layers and leachate channels. The process of continuous landfilling in the final disposal site (TPA) produces pollutants in the form of leachate (leachate) as a result of the infiltration of rainwater that enters the landfill. Research on the content of heavy metals in the leachate of TPA Batu Bola has never been carried out, therefore it is necessary to conduct research on the content of heavy metals in leachate at the Batu Bola landfill.

The purpose of this study was to determine the content of heavy metals Pb, Cu and Hg in the leachate of Batu Bola landfill. The benefit of this research is to provide initial information regarding the heavy metal content in the leachate of Batu Bola landfill so that it can be used as a reference material in controlling pollution.

2. Methods

The materials used are liquid waste media or leachate at Batu Bola landfill, Padangsidimpuan City, to be precise in Batu Bola Village, Padangsidimpuan Selatan District. The heavy metals used are Pb, Cu and Hg. The parameters tested were metal content or metal concentration. The equipment used consists of AAS, plastic sample bottles of 1.5 L size, label paper, stationery. The leachate samples were analyzed through a digestion process using AAS (Atomatic Absorption Spectrophotometer) at a wavelength of 283.3 nm. The liquid was analyzed to determine the content of heavy metals in the sample with the standard procedure of SNI 6989.8-2009 which was carried out at the Testing Laboratory of the Medan Industrial Research and Standardization Center.

The levels of heavy metals Pb, Cu and Hg in leachate obtained from the test results are then compared with the individual water quality standards set by the Government, namely through Permen LHK P.59 / 2016 and Permen LHK No.5 / 2014.

3. Results and Discussion

3.1 Heavy Metal Content in TPA Leachate Water

Leachate has certain characteristics, this is because the waste disposed of at the landfill comes from various different sources with different types of waste. The composition of leachate is not only influenced by the characteristics of waste (organic and inorganic), but also the ease of decomposition (soluble / insoluble), the condition of the waste pile (temperature, pH, humidity, and age), the characteristics of the water source (quantity and quality of water supplied). influenced by climate and hydrogeology), cover soil composition, nutrient availability, and microbes and the presence of inhibitors. Climate is an important factor affecting the quantity and quality of leachate. [9]. Becomes the transport phase for washing and migration of contaminants from the garbage heap, also affects biology. Likewise, the age of the pile of garbage also affects the quality of leachate and gas formed. Changes in the quality of leachate and gas are the main parameters to determine the level of stabilization of waste piles [10].The entry of chemical substances contained in leachate into the aquatic ecosystem can also affect the existing biota. If in the aquatic ecosystem there is pollution, it can cause the death of biota or affect physiological activities, cell formation and cell tissue function of an organ[11]. The variation in leachate composition is influenced by several factors, including: composition and age of waste, location and operation as well as landfill conditions, climate and hydrogeological conditions, humidity, temperature, pH, and level of stabilization [12].

Leachate has the characteristics of containing dissolved organic matter (CH4, voicyl fatty acid); inorganic macrocomponents (such as Ca^{+2}, Mg^{+2}, NH_4^+, Cl^-, $S04^{2-}$) heavy metals such as Pb, Cd, Cr, Cu, Ni, Hg, and others (San, 2001). Leachate from landfills has caused serious pollution. Organic compounds include aromatic hydrocarbons, phenols in groundwater, with very high concentrations of ammonium, heavy metals, and organic contaminants. In general, the main parameters used to describe the contaminant concentration in leachate are COD, ammonia. Nitrogen (NH4-N), Suspended Solids (SS), Dissolved Solids (DS), heavy metals, and salts [11].

The results of measuring the levels of Hg, Pb and Cu in leachate water at Batu Bola landfill show that the content of heavy metals Hg, Pb and Cu in leachate is still very low, data shows the detection of Hg, Pb and Cu metal concentrations is still below the AAS detection limit.

Table 1. Results of Measurement of Heavy Metals in Leachate Water at TPA Batu Bola

Parameter	Test results		
	Results	Unit	Methode
Mercury	<0,0004	mg/L	SNI 6989.8-2009
Lead	<0,003	mg/L	SNI 6989.8-2009
Copper	<0,006	mg/L	SNI 06-2462-1991

Measurement of mercury (Hg) levels in leachate using AAS is <0.0004 mg/L. The data shows that the Hg metal content in leachate is still below the threshold set by the government, which is 0.005 mg/L. The concentration of lead metal (Pb) also shows a value that is below the quality standard threshold, namely <0.003 mg/L, while the set limit is 0.1 mg/L. Likewise, the measurement of copper metal content (Cu) is still below the quality standard threshold, namely <0.006 mg/L, while the limit set by the government is 2.0 mg/L. The heavy metal content of copper probably comes from the waste of electrical equipment, pipes and wire.[13].

Table 2. Leachate Quality Standards

Parameter	Highest levels	
	Value	Unit
Mercury	0,005*	mg/L
Lead	0,1**	mg/L
Copper	2,0**	mg/L

Information :
* = Permen LHK P.59/2016 about the quality standards of leachate water
** = Permen LHK No.5/2014 regarding waste quality standards that have not been determined

Essential heavy metals are metals with certain concentrations needed by organisms, but these metals can cause toxic effects if excessive amounts such as the Cu concentration found in the leachate of Batu Bola landfill are classified as essential heavy metals. But you have to be careful because if it accumulates in large quantities it can still be dangerous. Non-essential heavy metals, metals whose benefits are not known in the body, are even toxic, such as Hg and Pb. Based on the results of the analysis, the metal concentration is still below the threshold set by the government, namely the Permen LHK P.59 / 2016 concerning the quality standard of leachate and / or final waste treatment activities. However, you also have to be careful because the heavy metals in these non-essential groups are toxic. Considering that in the future Indonesia is still faced with the problem of environmental pollution due to development, the efforts to restore and rehabilitate polluted land need mutual attention. At present, although phytoremidation technology has not been widely applied in restoring soil and water pollution, in the future it is hoped that it will become one of the potential environmental cleaning technologies supported by plant biodiversity in Indonesia that can be used as an accumulator boost plant, so that sustainable development programs can be achieved. [14].

4. Conclusions

Based on the measurement results of several parameters that have been carried out on the content of mercury, copper and lead in the leachate of Batu Bola landfill, it shows that the mercury, copper and lead content are still below the threshold set by the government. The data shows that the concentration of Hg metal is <0.0004 mg/L. Meanwhile, the threshold set by the government is 0.005 mg/L. The concentration of lead metal (Pb) also shows a value that is below the quality standard threshold, namely <0.003 mg/L, while the set limit is 0.1 mg/L. Likewise, the measurement of copper metal content (Cu) is still below the quality standard threshold, namely <0.006 mg/L, while the limit set by the government is 2.0 mg/L. However, caution must be exercised because if it is accumulated in large quantities it can still be dangerous.

References

[1] M. M. A. E.-S. and G. I. Abu-Zuid, “Imapact of Landfill Leachate on The Groundwater Quality: A Case Study in Egypt,” J. Adv. Res., vol. 6, pp. 579–586, 2015.
[2] S. dan P. S. Rakhmawati, E. A, “Kajian Pengelolaan Air Lindi (Leachate) di Lingkungan Tempat Pembuangan Akhir (TPA) Putri Cempo Surakarta Berbasis Kemanfaatan,” J. Ekosains, vol. 7, no. 4, pp. 8–15, 2016.
[3] S. dan Y. W. Warsinah, “Kajian Cemaran Logam Berat Timbal (Pb) Pada Kompartemen di Sekitar Tempat Pembuangan Akhir (TPA) Sukawinata Palembang,” J. Penelit. Sains, vol. 17, no. 2, pp. 78–81, 2015.
[4] and S. G. T. V. Anilkumar, A., D. Sukamaran, “Effect of Municipal Solid Waste Leachate on Ground Water Quality of Thiruvananthapuram District, Kerala, India,” Appl. Ecol. Environ. Sci., vol. 3, no. 5, pp. 151–157, 2015.
[5] and M. M. Fard, M. P, A. H. Mahvi, A. Asgari, “Heavy Metals Monitoring in Leachate from landfill Site of Qazvin, Iran,” Arch Hyg Sci, vol. 6, no. 1, pp. 44–48, 2017.
[6] dan E.-M. Al-Wabel, M. I., Al Yehya, W. I., Al-Farraj, S. E., “Characteristic of Landfill Leachates and Bio-Solids of Municipal Solid Waste (MSW) in Riyadh City Saudi Arabia,” J. Saudi Soc. Agric. Sci., vol. 10, p. 6570, 2011.
[7] T. Zubair, A., Malamassam, M., R., Syafitri, A., “Analisis Kualitas Air Lindi TPA Tamangapa Dan Pengaruhnya Terhadap Lingkungan,” J. Tek. Sipil Univ. Hasanuddin Makasar, pp. 4–8, 2015.
[8] W. Z. W. Radzuan, N. Z. M., Yaacob, “Samsudin, A. R., Rafek, A. G., 2005, Characteristic of Leachates at the Air Hitam Sanitary Landfill in Puchong Selangor,” Geol. Soc. Malaysia Bull., vol. 5, pp. 41–46, 2005.
[9] R. . Moelyaningrum, A.D dan Pujiati, “Cadmium (Cd) and Mercury (Hg) in the Soil, Leachete and Ground Water at the Final Waste Disposal PakusariJember District Area,” Int. J. Sci. Basic Appl. Res., vol. 24, no. 2, pp. 101–108, 2015.
[10] S. Kanmani and R. Gandhimathi, “Assessment of Heavy Metal Contamination in Soil Due to Leachate Migration from an Open Dumping Site,” Appl Water Sci, vol. 3, pp. 193–205, 2013.
[11] G. Christensen, T.H., Bjerg, P. 1,., Jensen. D. L., Heron, “Biogeochemistry of Landfill Leachate Plumes,” Appl. Geochem, vol. 16, p. 659, 2001.
[12] Tchobanolous, Solid Waste Engeneering Principles and Management Issues. New York: Mc Graw Hill Book Company, 1997.

[13] R. . dan A. Sari, "Karakteristik Air Lindi (Leachate) di Tempat Pembuangan Akhir Sampah Air Dingin Kota Padang," J. Fis. Unand, vol. 6, no. 1, pp. 93–99, 2017.
[14] dkk Irhamni, Kandungan Logam Berat pada Air Lindi Tempat Pembuangan Akhir (TPA) Sampah Kota Banda Aceh. Banda Aceh, 2017.

Acknowledgments
Thank you to the Chancellor UniversitasMuhammadiyahTapanuli Selatan . This research was supported and funded by UMTS through an internal grant 2020.

IOP Conf. Series: Materials Science and Engineering 1156 (2021) 012013 doi:10.1088/1757-899X/1156/1/012013

Biogas : Renewable Energy

Lelya Hilda*, **Rosimah Lubis and Replita**

Institut Agama Islam Negeri Padangsidimpuan, Indonesia.

* lelya@iain-padangsidimpuan.ac.id

Abstract. Biogas has several advantages comparing on the use of fossils fuels such as gasoline and natural gas. Biogas can be produced continuously because it grows from the organic raw materials to create a clean and healthy environment and free pollution. Utilization of animal waste as an energy source of biogas will not reduce the amount of organic fertilizer produced, but will produce methane (CH4) as an energy source. The research was done by mixing cow dung, rumen and water with various comparing materials and time to produce the highest biodiesel. The result of research by mixing cow dung (rumen) and water produced the highest biogas in a 1: 2 mixture. From the 3rd day, it has produced biogas but cannot be used because the CO and CO2 contents are very high. The result on the 10th day, it can be already used. The highest biogas production was resulted on 15th – 16th day in which producing 5 liters biogas from 110 kg cow waste with 220 kg water contents. The results are getting smaller but it still produces until the 30th day. The highest result to produce renewable energy was by utilizing without reducing the function of organic fertilizer.

1. Introduction

Indonesia is one of the producers of fuel and gas but human activities using fuel are increasing while the supply of these fuels is running low. The United Nations (UN) states that the use of fossil fuels such as petroleum, natural gas and coal is a carbon dioxide contributor to the earth's atmosphere which causes the greenhouse effect and influences of global warming.

Energy is an important component to support productive activities and businesses in producing goods and services[1],[2]. Energy sources can come from fossil energy, solar energy, water, wind or energy from biological resources (bio energy). The scarcity of fuel is inevitable, especially petroleum, but the depletion of supplies has caused prices to soar[3],[4, p. 34].

Energy serves as a driver of national development both in industrial activities, transportation, services and households[5],[6],[7]. The use of fossil fuels that cause negative effects on the environment must be found out the alternatives to reduce the effect caused. The increasing of fuel needs and limited supplies force us to look for other alternatives to substitute fuels by renewable energy sources that is renewable energy used to replace fossil fuels.

New innovations have been found in overcoming critical energy that is cheaper and can be reached by the community and environmentally friendly. One of alternatives renewable energy is biogas which is from the fermentation process of organic waste materials such as livestock manure, organic waste and other materials by methanogenic bacteria in anaerobic conditions that was without oxygen[8],[4, p. 8].

Biogas is a renewal energy source as a fuel contributor that is needed by the community. This biogas is a non-fossil fuel, which is waste or manure or other manure. Whereas natural gas as renewal energy derived from fossils whose formation takes millions of years [9]. Biodiesel as an alternative diesel fuel is described as methyl fatty acids or ethyl esters from vegetable oils or animal fats[10].

Biodiesel is a renewable, one of the environmentally friendly alternative fuel produced from renewable sources such as vegetable oils and animal fats. Vegetable oils, animal fats and used oils are reacted with alcohol through a transesterification reaction[11].

Utilization of livestock waste from cattle dung is one appropriate alternative to overcome the rising price of fertilizers and scarcity of fuel oil, namely making biogas. Utilization of livestock manure as an energy source (biogas), will not reduce the amount of organic fertilizer producing. Producing biogas from livestock manure produces methane (CH_4) as an energy source and then returned to its original condition. Livestock manure that has been processed for making biogas is moved to a drier place, and when it is dry, it is stored in a sack for future using[12].

The development of biogas is one of overcoming energy critical. It is time for Indonesia to reduce its dependence on fuel oil by developing alternative energy sources that are environmentally friendly and renewable (renewable). One alternative fuel is bioenergy. Bioenergy is suitable for overcoming energy problems due to several advantages. Bioenergy is renewable, environmentally friendly, and biodegradable, reducing greenhouse effect and continuity of raw materials is quite guaranteed. Bioenergy raw materials can be obtained in a simple way through the cultivation of biofuel-producing plants and utilizing waste around human life [13].

In line with the background, the purpose of this research is to produce biogas from cow dung as a renewable energy source. Sustaining waste treatment will reduce water, soil and air pollution by processing it into biogas and organic fertilizer.

2. Methods

Biogas isproducedfromorganicwaste. The materials needed are cow dung, cowstomach rumen liquid, andwater, while the equipment use displastic bottles, measuring cups, plastichoses, rubber bottle caps, clampclips, plastic water boxes, wire, stative, and clamps. For the making of biogas, prior research will be carried out, bymixingcow dung, rumen and water with various comparisons. The high estratio producing biogas will be used for further biogas production

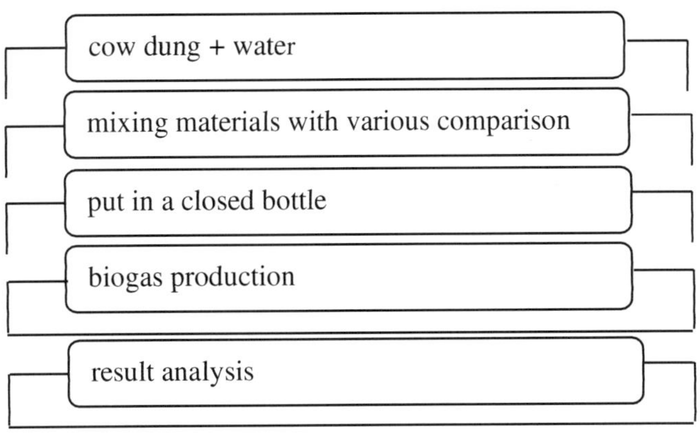

Figure 1. Biogas Production Process Scheme

3. Results and Discussion

The result of this research shows that the producing of biogas-making composition mixture and length of time has produced maximum biogas. Biogas was produced by mixing cow manure and water in producing the highest biogas in 1: 3 mixtures. From day 3 it has produced biogas but cannot be used because the CO and CO_2 contents are very high, results on day 10 can already be used and the highest biogas production on days 15-16 produces 5 liters of biogas from 110 kg of cow dung with a water content of 220 kg. The results are getting smaller but still producing until the 30th day.

From the producing result, the highest obtained at a ratio of 1: 3, but the high contents of CO and CO2 causes it cannot burn properly. The second result in ratio1: 2 also produce the highest biogas but CO and CO2 contents are low so that the process of combustion is good. The last result in ratio 1:1 produce maximum biogas by taking longer time and resulted was fail because the bacteria die in too dense. The complete results can be seen in Table 1 and Figure 1 below:

Table 1. Ratio and Result

Day	**1 :1**	**1 : 2**	**1 : 3**
1	0.2	0.1	0.1
2	0.3	0.3	0.5
3	0.5	0.6	1.4
4	0.7	0.9	1.8
5	1	1.4	2.5
6	1.2	1.7	2.9
7	1.3	1.9	3.6
8	1.4	2.4	4.2
9	1.6	2.7	5.1
10	2.1	3.7	5.9
11	2.4	4.1	6.2
12	2.7	4.3	6.1
13	2.8	4.5	5.7
14	3	4.7	5.5
15	3.2	5.1	5.1
16	3.4	5.2	4.2
17	3.6	5	4
18	3.8	4.8	3.8
19	3.9	4.5	3.6
20	4.2	4.1	3.4
21	4	3.6	2.9
22	3.6	3.4	2.5
23	3.2	3.1	2.2
24	2.6	2.8	1.9
25	2.3	2.7	1.5
26	1.9	2.4	1.4
27	1.3	2.2	1.2
28	0.6	1.4	0.6
29	0.2	0.9	0.4
30	0.1	0.3	0.2

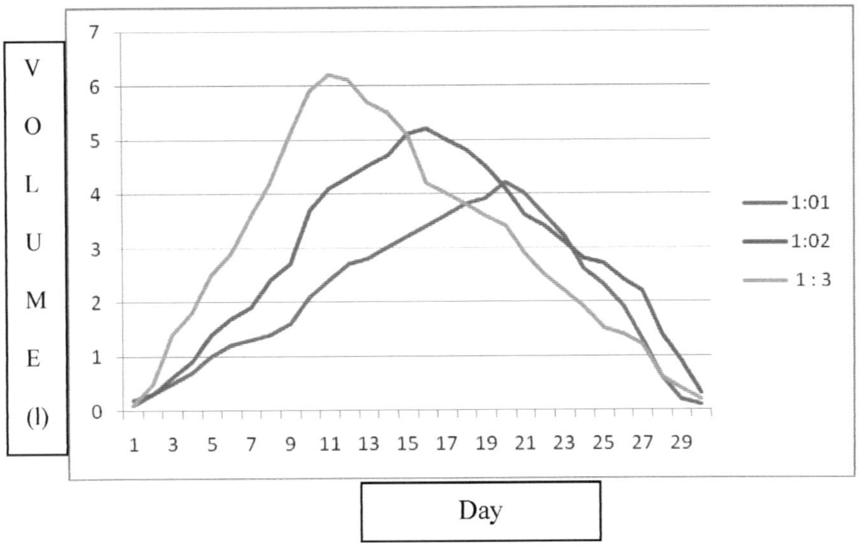

Figure 2. Biogas Production Process

Waste from cow dung can be used as biodiesel toreduce green house gas emissions and to facilitate the development of a sustainable energy supply. Production of biogas produces renewable energy that is versatile, because me thane can be use dtoreplace fossil fuels both in electricity and heat generation and as vehicle fuel. In creasing and utilizing gas biogas as a renewable vehicle fuel orinjection in to the natural gas network is in creasingly attractive because gas can be used more efficiently[14].

Energy balance depends on biogas production, utilization efficiency, and the value of energy substitutes for fossil fuels. Poschletal's[15] research showed that in creasing biogas to biomethane for injection in to natural gas networks has the potential to in crease primary energy in put for biogas utilization by up to 100%; the energy efficiency of the biogas system in creases to 65% when natural gas is replace dinstead of electricity. It was also found that, the energy efficiency of the system could be further in creased by 5.1-6.1% through the restoration of there maining biogas from a closed digestate storage unit.

Generally, in creasing of water content will in crease biogas production. Water molecules are needed during the stages of hydrolysis and acetogenesis. In the hydrolysis stage, by helping of hydrolytic microbes degrade complex organic compounds in the form of polymer sintomonomers in the form of in soluble compounds with ligh termolecular weights. At the stage of acetogenesis is the stage where the results of the previous stage, acidogenesis are converted to acetic acid. Ethanol, propionateacid, and butyric acid are converted to aceticacid by the help of acetogenic bacteria.

In the process of converting organic acid stoacetic acid excesses water molecules are needed, so the addition of more water willin crease the formation of acetic acid which will later be converted into methane gas at a later stage (methanogenesis). Thus the addition of more water in the system, biogas production will increase[16].

During the first 3 day sofobservation, biogas production was reduce dand mainly due to the lagging phase of microbialgrowth. In fact, in the range of 4 to 6 day sofobservation; biogas production in creases substantially for exponential methanogengrowth. The highest biogas production rate of 3.4 L was measured on the 6th day. On the date of the start of semi-sustainable digestion, biogas production was observed to decrease significantly and, this maybe because an unregulate dpH area wasused, which simultaneouslyled to an increase in possible ammonia nitrogen concentration considered to inhibit the process[17][18],[19].

Ukpai'sandNnabuchi,s research[20] shows that making biogas from cow dung with water produces the highest cumulative biogas of 124.3 L / total pulp mass (TMS). Powers of Hydrogen (pH) changes indicate that at neutral pH, the high estpeakof gas production is reached and in the slightly acid icp

Hrange, there is no gas production. These results indicate that this waste can become a renewable gas source ifitis manage dproperly because each waste is slow in continuing gas production after 30 day sofretention time.

The research of Saputro et al[16] shows that the high estratio of animal manure with wateris 1: 3, but this research has not seen which is better in the combustion process. In the hydrolysis stage, with the help of hydrolytic microbes degrade complex organic compounds in the formof polymersintomonomers in the formo finsoluble compounds with lighter molecular weights. At the stage of acetogenesis is the stage where the results of the previous stage, acidogenesis, are converted to acetic acid. Ethanol, propionate acid, and butyric acid are converted to acetic acid by the help of acetogenic bacteria. In the process of converting organic acid stoacetic acid, excess water molecules are needed, so the addition of more water will in crease the formation of acetic acid which will later beconverted into methane gas at a later stage (methanogenesis). Thus the addition of more water in the system, biogas production will increase.

4. Conclusions

Biogas production process during 30 days have resulted maximum biogas. The results are getting smaller but still producing until the 30th day. Biogas as renewable energy sources are used to create bio-energy can greatly contribute to reducing greenhouse gases as they have the possibility of reducing the need to use fossil fuels. By providing a non-polluting energy source which is also renewable, the earth is being kept clean of harmful emissions.

References

[1] J. Terrapon-Pfaff, M.-C. Gröne, C. Dienst, and W. Ortiz, "Productive use of energy – Pathway to development? Reviewing the outcomes and impacts of small-scale energy projects in the global south," *Renewable and Sustainable Energy Reviews*, vol. 96, pp. 198–209, Nov. 2018, doi: 10.1016/j.rser.2018.07.016.

[2] V. S. Arutyunov and G. V. Lisichkin, "Energy resources of the 21st century: problems and forecasts. Can renewable energy sources replace fossil fuels†," *Russ. Chem. Rev.*, vol. 86, no. 8, p. 777, Aug. 2017, doi: 10.1070/RCR4723.

[3] K. T. Sanders and M. E. Webber, "Evaluating the energy consumed for water use in the United States," *Environ. Res. Lett.*, vol. 7, no. 3, p. 034034, Sep. 2012, doi: 10.1088/1748-9326/7/3/034034.

[4] S. Wahyuni, *Biogas: Energi Alternatif Pengganti BBM, Gas dan Listrik*. Jakarta: AgroMedia Pustaka, 2013.

[5] K. Kaygusuz, "Energy services and energy poverty for sustainable rural development," *Renewable and Sustainable Energy Reviews*, vol. 15, no. 2, pp. 936–947, 2011, doi: 10.1016/j.rser.2010.11.003.

[6] M. A. Brown, A. Soni, M. V. Lapsa, K. Southworth, and M. Cox, "High energy burden and low-income energy affordability: conclusions from a literature review," *Prog. Energy*, vol. 2, no. 4, p. 042003, Oct. 2020, doi: 10.1088/2516-1083/abb954.

[7] L. Adua and B. Clark, "Even for the environment, context matters! States, households, and residential energy consumption," *Environ. Res. Lett.*, vol. 14, no. 6, p. 064008, 2019, doi: 10.1088/1748-9326/ab1abf.

[8] M. G. Mengistu, B. Simane, G. Eshete, and T. S. Workneh, "Factors affecting households' decisions in biogas technology adoption, the case of Ofla and Mecha Districts, northern Ethiopia," *Renewable Energy*, vol. 93, pp. 215–227, 2016, doi: 10.1016/j.renene.2016.02.066.

[9] T. Hariati, "Biogas: Limbah Peternakan yang Menjadi Sumber Energi Alternatif," *Jurnal Wartazoa*, vol. 16, no. 3, pp. 160–169, 2016, doi: http://medpub.litbang.pertanian.go.id/index.php/wartazoa/article/download/858/867.

[10] K. Sivaramakrishnan and P. Ravikumar, "Determination of Higher Heating Value of Biodiesels," *International Journal of Engineering Science and Technology*, vol. 3, no. 11, pp. 1981–1987, 2011, doi: https://www.idc-online.com/technical_references/pdfs/electrical_engineering/DETERMINATION%20OF%20HIGHER%20HEATING%20VALUE%20OF%20BIODIESELS.pdf.
[11] M. Faizal, U. Maftuchah, and W. A. Auriyani, "Pengaruh Kadar Metanol, Jumlah Katalis, dan Waktu Reaksi Pada Pembuatan Biodiesel dari Lemak Sapi Melalui Proses Transesterifikasi," *Jurnal Teknik Kimia*, vol. 19, no. 4, pp. 29–37, 2016, doi: http://jtk.unsri.ac.id/index.php/jtk/article/download/155/154.
[12] S. Rahayu, D. Purwaningsih, and Pujianto, "Pemanfaatan Kotoran Ternak Sapi sebagai Sumber Energi Alternatif Ramah Lingkungan Beserta Asfek Sosio Kulturalnya," *INOTEKS*, vol. 13, no. 2, Art. no. 2, 2019, doi: https://journal.uny.ac.id/index.php/inotek/article/view/38.
[13] W. W. Widaningsih, "Partisipasi Masyarakat Melalui Desa Mandiri Energi Berbasis Biogas Limbah Ternak Sapi di Desa Haurngombong Kecamatan Pamulihan Kabupaten Sumedang," *Jurnal Ilmu Administrasi: Media Pengembangan Ilmu dan Praktek Administrasi*, vol. 11, no. 1, Art. no. 1, Apr. 2014, doi: 10.31113/jia.v11i1.26.
[14] P. Weiland, "Biogas production: current state and perspectives," *Appl Microbiol Biotechnol*, vol. 85, no. 4, pp. 849–860, Jan. 2010, doi: https://link.springer.com/article/10.1007/s00253-009-2246-7.
[15] M. Poschl, S. Ward, and P. Owende, "Evaluation of energy efficiency of various biogas production and utilization pathways," *Applied Energy*, vol. 87, no. 11, pp. 3305–3321, Nov. 2010, doi: http://www.sciencedirect.com/science/article/pii/S0306261910001790.
[16] R. R. Saputro and Rr. D. A. Putri, *Pembuatan Biogas dari Limbah Peternakan*. Semarang: Department of Chemical Engineering Universitas Diponegoro, 2009.
[17] S. Hidayati, T. P. Utomo, E. Suroso, and Z. A. Maktub, "Technical and technology aspect assessment of biogas agroindustry from cow manure: case study on cattle livestock industry in South Lampung District," *IOP Conf. Ser.: Earth Environ. Sci.*, vol. 230, p. 012072, Feb. 2019, doi: 10.1088/1755-1315/230/1/012072.
[18] A. K. Jha, J. Li, Q. Ban, L. Zhang, and B. Zhao, "Dry Anaerobic Digestion of Cow Dung for Methane Production: Effect of Mixing," *Pakistan Journal of Biological Sciences*, vol. 15, no. 23, pp. 1111–1118, 2012, doi: https://dx.doi.org/10.3923/pjbs.2012.1111.1118.
[19] B. S. U. I. Abubakar and N. Ismail, "Anaerobic Digestion of Cow Dung for BiogasProduction," *ARPN Journal of Engineering and Applied Sciences*, vol. 7, no. 2, pp. 169–172, 2012, doi: https://www.researchgate.net/profile/Baba_Shehu_Umar_Ibn_Abubakar/publication/234162510_Anaerobic_digestion_of_cow_dung_for_biogas_production/links/09e4150fa3f59d9e31000000.pdf.
[20] P. A. Ukpai and M. N. Nnabuchi, "Comparative Study of Biogas Production From Cow Dung, Cow Pea and Cassava Peeling Using 45 Litres Biogas Digester," *Advances in Applied Science Research*, vol. 3, no. 3, pp. 1864–1869, 2012, doi: https://www.imedpub.com/abstract/comparative-study-of-biogas-production-from-cow-dung-cow-pea-and-cassavarnpeeling-using-45-litres-biogas-digester-16356.html.

Acknowledgments
The author expressed his gratitude to the IAIN Padangsidimpuan on Applied Research of the National Development through a BOPTN Research grant of research and community service in 2019.

Effect of variations in Composition and Sintering Temprature on Mechanical Properties Ceramic Membrane with Bentonite, Sinabung Volcanic Ash and Carbon Active

Moraida Hasanah[1,2*], Timbangen Sembiring[1], Kerista Sebayang[1], Syahrul Humaidi[1], Rahmadsyah[2] and Fynnisa Z[3]

[1]Department of Physics, Faculty of Mathematics and Natural Science, Universitas Sumatera Utara, Medan, Indonesia.
[2]Department of Mechanical Engineering Department, University of Asahan, Asahan, Indonesia.
[3]Department of Civil Engineering Department, University of Asahan, Asahan, Indonesia.

* hasanahmoraida@gmail.com

Abstract. All articles must contain an abstract.The abstract text should be formatted using 10 point Times or Times New Roman and indented 25 mm from the left margin. Leave 10 mm space after the abstract before you begin the main text of your article, starting on the same page as the abstract. The abstract should give readers concise information about the content of the article and indicate the main results obtained and conclusions drawn. The abstract is not part of the text and should be complete in itself; no table numbers, figure numbers, references or displayed mathematical expressions should be included. It should be suitable for direct inclusion in abstracting services and should not normally exceed 200 words in a single paragraph. Since contemporary information-retrieval systems rely heavily on the content of titles and abstracts to identify relevant articles in literature searches, great care should be taken in constructing both.

1. Introduction

Ceramic material is a non-metallic material which is formed by a heating process. Now ceramics are widely used as electronic components, kitchen utensils, building materials and as a means of filtering water, for example ceramic membranes [3].

Membrane ceramic is a material that is categorized as an inorganic membrane. Ceramic membrane has an advantage over other membranes because they are resistant to heat, acids and alkalis. The raw materials for making ceramics are bentonite, silica, clay, quartz and alumina, while the pore-forming materials other than inorganic materials such as carbon and calcium carbonate, are also organic materials such as rice flour, activated carbon, potatoes and corn [6].

Kaolin, clay, feldspar and quartz and bentonite. Natural bentonite is a hydrated silicate alumina with the main element consisting of alkaline and alkaline earth cations from the compounds it contains. Bentonite means clay which contains hydrate compounds of alumino silicates with the main elements of alkaline soil and has ion exchange properties and high absorption ability [2]. In addition, one of the materials that can be used is the volcanic ash of Mount Sinabung. Volcanic ash is magma fragments and consists of minerals, volcanic glass and also contains materials high in silica and aluminum [7]. The content of volcanic ash was taken from Mt. The eruption of Mount Sinabung from the Berastepu Kec Village area. Simpang Empat Kab. All type volcanic eruptions are capable of causing dangerous and destructive phenomena [8]. However, volcanic ash in Karo and is used as a raw material in the manufacture of paving block [10] or ceramic membranes because volcanic ash compounds contain SiO2 (78.3%), Fe2O3 (2.91%), Al2O3 (4.56%), MgO (1.07) compounds. %), CaO (4.84%), Na2O (0.46%) [4].

Research that has been done using volcanic ash material, namely porous ceramics for motor vehicle exhaust gas filters and catalytic converters. This research tries to utilize the volcanic ash material of Mount Sinabung for ceramic membranes with carbon active [9] [11].

2. Methods

In this study, the manufacture of ceramic membranes using bentonite, Sinabung volcanic ash and activated carbon using conventional techniques with variations in the composition of bentonite, Sinabung volcanic ash and activated carbon are presented in Table 1.

Table 1. Comparison of Porous Ceramic Material Composition

Bentonite (%wt)	Volcanic Ash (%wt)	Carbon Active (%wt)
80	**0**	**20**
75	**5**	**20**
70	**10**	**20**
65	**15**	**20**
60	**20**	**20**

The three ingredients are mixed evenly and then printed using a pellet mold. After the ceramic membrane samples were printed in the form of pellets, then each sample was burned with a temperature treatment of 800°C, 900°C and 1000°C and for 3 hours.

After the ceramic membrane sample in the form of a pellet was sintered, then each sample was characterized for its mechanical properties, namely the compressive test and hardness test using the UTM and Hardness Testing equipment available at the PTKI Medan Material Test Laboratory.

3. Results and Discussion

3.1 Compressive strength

The compressive strength value of ceramic membrane shows that there is an effect of sintering temperature on the compressive strength of ceramic membranes made of bentonite, volcanic ash and activated carbon as shown in Table 2.

Table 2. Compressive Strength Ceramic Membran

Bentonite: Volcanic Ash: Carbon Active	**Compressive strength (MPa)**		
	800° C	**900° C**	**1000° C**
80:00:20	2.6	2.96	3.02
75:5:20	2.8	3.03	3.25
70:10:20	3.2	4.5	5.41
65:15:20	4.3	4.7	6.66
60:20:20	4.5	4.75	5.19

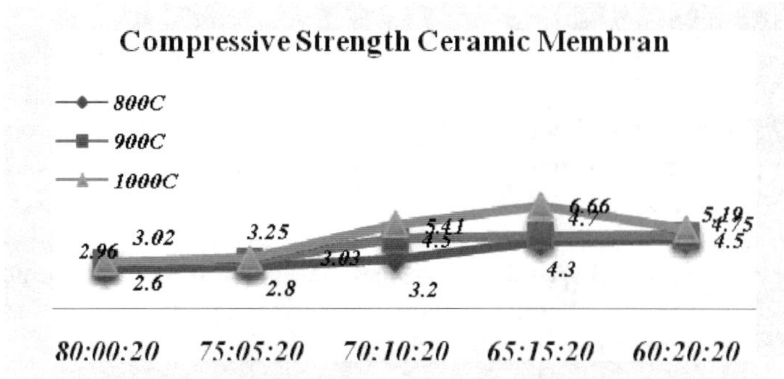

Figure 1. Graph of Compressive Strength with composition of Bentonite,Volcanic Ash and Carbon Active

The results showed that the optimum composition of bentonite, Sinabung volcanic ash and activated carbon at the composition of 65: 15: 20 with a temperature treatment of 1000°C had the highest compressive strength value of 6.6 MPa. Almost all variations mix, there is a relationship directly proportional to the sintering temperature and the compressive strength of the ceramic. The greater the sintering temperature, the more large compressive strength of the ceramic membrane generated.

3.2 Hardness

The hardness value of ceramic membrane shows that there is an effect of sintering temperature on the hardness of ceramic membranes made of bentonite, volcanic ash and activated carbon as shown in Table 3.

Table 3. Hardness Ceramic Membran

Bentonite: Volcanic Ash: Carbon Active	Hardness (MPa)		
	800° C	**900° C**	**1000° C**
80:00:20	93.320	108.220	117.112
75:5:20	106.540	117.340	149.859
70:10:20	118.430	130.671	153.428
65:15:20	175.121	182.422	187.639
60:20:20	132.826	102.212	142.826

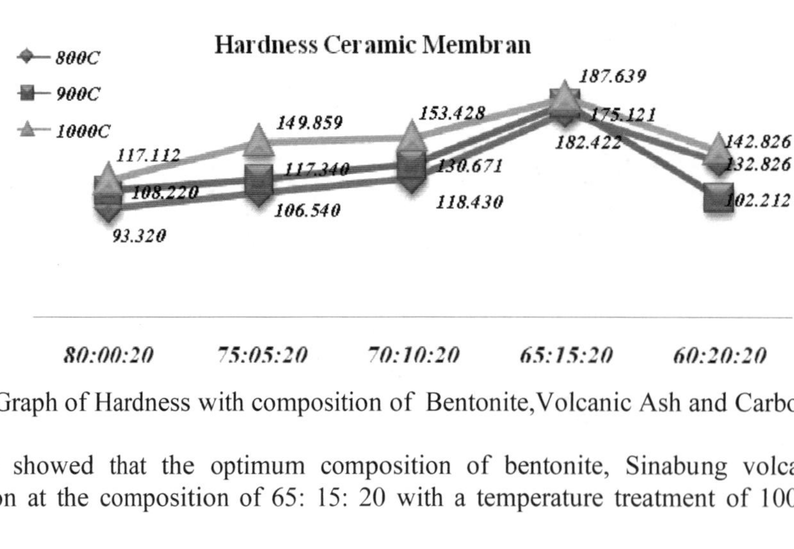

Figure 2. Graph of Hardness with composition of Bentonite,Volcanic Ash and Carbon Active

The results showed that the optimum composition of bentonite, Sinabung volcanic ash and activated carbon at the composition of 65: 15: 20 with a temperature treatment of 1000oC had the highest hardness of 187.639MPa. This shows that the greater the addition of volcanic ash to the composition of the ceramic membrane, the higher hardness values of the ceramic membrane. Likewise with the sintering temperature, the more the value of the ceramic membrane combustion temperature, the higher the value of mechanical strength, namely compressive strength and hardness [1].

4. Conclusions

This research presents the effect of variations in the composition of bentonite, volcanic ash from Mount Sinabung and activated carbon on the mechanical properties of membranes. This study also affects the effect of ceramic membrane sintering temperature on the mechanical properties of ceramic membranes. Based on the results of the research, the increase in compressive strength and hardness values was caused by the addition of volcanic ash to the composition (60:15:20) and an increase in temperature at sintering1000oC had an optimum compressive strength value of 6.6 MPa and hardness of 187.639MPa.

References

[1] A.(Ed.). Lakshmanan, *Sintering of ceramics: new emerging techniques* (BoD–Books on Demand, 2012).

[2] Bachtiar, Iyas Muzalif, Sudarsono Sudarsono, and Abd Kadir. "Pengaruh Fraksi Volume Pasir Terhadap Kekuatan Bending, Densitas Dan Porositas Keramik Berbahan Dasar Tanah Liat." *ENTHALPY* 4.3, 2019.

[3] Haryanti, Ninis Hadi, and Tetti Novalina Manik. "PENGARUH SUHU SINTERING TERHADAP SIFAT MEKANIK KERAMIK BERBAHAN LEMPUNG DAN ABU SEKAM PADI." *Jurnal Fisika Flux: Jurnal Ilmiah Fisika FMIPA Universitas Lambung Mangkurat* 13, no. 1 pp: 1-10, 2016.

[4] Hasanah, Moraida, et al. "Effect of the mass and volume shrinkage of porous ceramics on Sinabung volcanic ash." *AIP Conference Proceedings*. Vol. 2221. No. 1. AIP Publishing LLC, 2020.

[5] Hasanah, M., & Fynnisa, Z. (2019, December). ANALISIS SIFAT EMISI GAS CO, HC, CO2 DAN O2 PADA MEMBRAN KERAMIK BERBASIS BENTONIT DAN DEBU VULKANIK GUNUNG SINABUNG. In *Seminar Nasional Multi Disiplin Ilmu Universitas Asahan*.

[6] Manohar, 2012. Development and Characterization of Ceramic Membrane, International Journal of Modern Engineering Research (IJMER). 2: 1492-1506.

[7] M. F. Serra, M. S. Conconi, G. Suarez, E. F. Aglietti, and N. M. Rendtorff, Ceram. Int. 41, 6169–6177, 2015.

[8] Nurwihastuti, Dwi Wahyuni, et al. "Volcanic hazard analysis of sinabung volcano eruption in karo north sumatra indonesia." *Journal of Physics: Conference Series*. Vol. 1175. No. 1. IOP Publishing, 2019.

[9] Saktisahdan, Tengku Jukdin, Moraida Hasanah, and Supriono Siregar. "PENGARUH KOMPOSISI TERHADAP SIFAT MEKANIK KERAMIK BERPORI MENGGUNAKAN DEBU VULKANIK GUNUNG SINABUNG."

[10] Sembiring, I. S., and I. P. Hastuty. "Sianbung Volcanic Ash utilization As The Additive for Papving Block Quality A and B." *IOP conf. Materials Science and Engeeneering*, 2017.

[11] Sinuhaji, P., Sembiring, T., & Handoko, F. Using Alumina and Volcanic Ash in Producing Catalytic Converter Ceramics. *Int. J. Appl. Eng. Res*, *12*, 343-347, 2017.

Acknowledgments

We would like thank to Direktur Riset dan Pengabdian Masyarakat (DPRM) Kementrian Riset, Teknologi dan Pendidikan Tinggi atas pembiayaan penelitian, Universitas Sumatera Utara and Universitas Asahan.

Extraction Of Silica Dioxide (SiO_2) From Mount Sinabung Volcanic Ash with Coprecipitation Method

Moraida Hasanah[1,2], Timbangen Sembiring[1*], Kerista Sebayang[1], Syahrul Humaidi[1], Rahmadsyah[2], Tengku Jukdin Saktisahdan[2], Fauzi Handoko[1] and Sally Irvina Ritonga[3]

[1]Department of Physics, Faculty of Mathematics and Natural Science, Universitas Sumatera Utara, Medan, Indonesia.
[2]Department of Mechanical Engineering Department, University of Asahan, Asahan, Indonesia.
[3]Integrated Research Laboratory, Universitas Sumatera Utara, Medan, Indonesia.

* timbangen@usu.ac.id

Abstract. Mount Sinabung volcanic ash is a basic material that can be used to produce silicon dioxide (SiO_2). SiO_2 can be used as a building raw material as well as in food and beverage packaging. To produce SiO_2, the volcanic ash of Mount Sinabung must be synthesized using the coprecipitation method. The result of this process produces SiO_2 content as much as 48%, a formed function group that states the existence of SiO_2 namely Si-O-Si and Si-O at absorption of 1095.57 cm-1 and 798.53 cm-1. From the results of SEM can be seen the morphology of SiO_2 formed. In the XRD results, the phase of SiO_2 is amorphous.

1. Introduction

Indonesia is an archipelago which is crossed by the equator and located between the continents of Asia and Australia and between the Pacific Ocean and the Indian Ocean. It is surrounded by the Eurasian, the Indo-Australian and the Pacific plate. In addition, this region is also the path of the Pacific Ring of Fire (Ring of Fire), which is a series of lines of active volcanoes in the world[1]. Mount Sinabung, which is called *Deleng Sinabung* in Karonese, one of active volcanoes in Indonesia. Located in the province of North Sumatra withcoordinates the summit of Mount Sinabung is 03° 10' N and 98° 23' E with the highest peak 2,460 meters above sea level the highest peak in North Sumatra[2][3]. Mount Sinabung eruption increased to the level of 4 (Alert) on November 24, 2013 since it spat out black, thick smokescreen, followed by sand rain and volcanic ash which covered thousands of hectares of farmers' crops under the radius of six kilometers so that more than 20 thousand people had to be evacuated[4].

The abundance of volcanic ash material as a result of the eruption of Mount Sinabung is an interesting thing to study. According to tests that have been done using Spektroskopi Serapan Atom (Atomic Absorption Spectrophotometry – AAS) showed that the volcanic ash from Mount Sinabung contains 78,3 % SiO_2 , 2,91 % Fe_2O_3 , 4,56 % Al_2O_3 , 1,07 % MgO, 4,84 % CaO, and 0,46 % Na_2O[5]. SiO2 is one of the most abundant compound in the Volcanic Ash Sinabung has the potential to be used a variety of needs[6]. One of the ways used to obtain high-purity SiO2 can be obtained by synthesis of SiO2-based nature with relatively high Si element content using coprecipitation method.The coprecipitation (deposition) method is a type of fabrication technique using a chemical method that brings a solute downward to form the desired precipitate[7][8][9].

Research that has been done using volcanic ash material , namely extraction volcanic ash of sinabung mount silica to production silica gel. This research tries extraction volcanic ash of sinabung monnt with sol-gel method[10].

2. Methods

In this study, the manufacture of extraction silica dioxide (SiO_2) using Volcanic ash of Sinabung mount ,Cholride acid (HCl) 2 M, distilled water , Natrium Chloride and Natrium Hidroxide (NaOH) 7M. This research using coprecipitation method.

Mount Sinabung volcanic ash was sieved with a 200 mesh filter.Put it in a container of 300 grams of Mount Sinabung volcanic ash that has been sifted.Add HCl 2 M with a ratio (1:10) and stirred for 12 hours.Settling for one night until the ash settles, the precipitate was washed with distilled water until pH = 7.Add 7 M NaOH (1:10);Then heated at a temperature of 100-130°C while in stirrer for 24 h. After reaction was complete, allowed to stand to separate.Furthermore, the filtrate is filtered with a paper filter Results whattman No. 42 in tittrasi with 2 M HCl until pH neutral.Settling up a separate deposition. Dried sludge by using an oven at 70-80°C to dry.Precipitated silica dioxide obtained weighing 68.58 grams of 300 grams of volcanic ash Mount Sinabung

3. Results and Discussion

3.1 XRF analysis of SiO_2 from Sinabung Volcanic Ash

The Mount Sinabung volcanic ash content has been extracted through the coprecipitation process can be seen in table 1. Based on the results of XRF shows the content of SiO_2 compounds as much as 48%, the content of elements Si 26.9 %, and the content of elements Cl 22.8 %. From the test results showed the content of SiO_2 has a fairly high content. It can also be observed with discoloration occurring. The Mount Sinabung volcanic ash color is gray, after the coprecipitation process turns white[11].

Table 1. Composition of SiO_2 from Sinabung Volcanic Ash used XRF

Composition	Percentage (%)
SiO_2	48
Si	26.9
Cl	22.8
P_2O_5	0.5
K_2O	0.17
CaO	0.66
Cr_2O_3	0.058
Fe_2O_3	0.61
BaO	0.28

SiO_2 content can be used in the field of industry, among others cement industry, ceramic industry, ferosillicone manufacturing industry. SiO_2 can not stand alone so it must bind to oxygen to form a compound that is silicon dioxide compound (SiO_2). So if the oxide level is getting smaller then it is expected that SiO_2 levels will be greater anyway. This means that the purer SiO_2 levels, so that it can be used in various fields. Examples of food and beverage packaging such as bottles and glasses[12].

3.2 Morphology analysis of SiO_2 from Sinabung Volcanic Ash

SiO_2 morpology from Mount Sinabung volcanic ash was observed using Scanning Electron Microscope (SEM). Test results using SEM can be observed in figure 1.

IOP Conf. Series: Materials Science and Engineering 1156 (2021) 012015 doi:10.1088/1757-899X/1156/1/012015

Figure 1. SEM image of SiO_2 from Sinabung Volcanic Ash.

In figure 1, the surface structure of SiO_2 is seen. SiO_2 shape looks like granules that appear that the surface has not seen the presence of grain boundry, with the size of pores large enough to form unevenly distributed blobs. These results indicate that SiO_2 has low stability[13]. Then from the results SEM seen other minerals that still cover the surface of SiO_2. It can also be linked to the XRF test hail which shows the still high content of Cl in the process to obtain SiO_2 from the Mount Sinabung volcanic ash.

3.3 FTIR Analysis

To find out the chemical bonds present in SiO_2 using FTIR. The FTIR test results for SiO_2 can be seen in figure 2. There are seven absorption tapes from the FTIR test results, the results of which are presented in table 2.

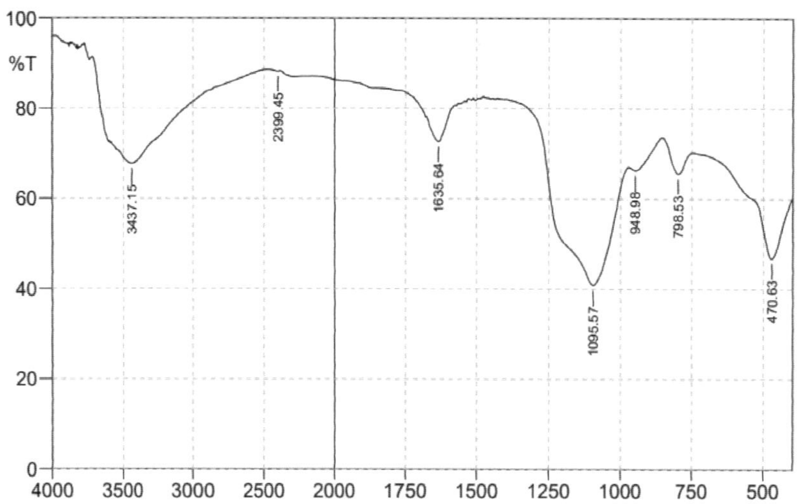

Figure 2. FTIR spectrum of SiO_2 from Sinabung Volcanic Ash

Bonds that appear on wave numbers 470.63 cm^{-1}, 798.53 cm^{-1}, 948.98 cm^{-1}, 1095.57 cm^{-1}, 1635.64 cm^{-1}, and 2399.45 cm^{-1}. The absorption band at 470.63 cm^{-1} is a bond of O-Si-O from siloxy group. Absorption at 1635.64 cm^{-1} and 3437.15 cm^{-1} is a hydroxyl function group of-OH stretching and -OH bending [14]. At absorption 798.53 cm^{-1} ,948.98 cm^{-1} and 2399.45 cm^{-1} is a hydroxyl function group of Si-O, Si-OH dan Si-C[15]. At absorption 1095.57 cm^{-1} is a siloxan function group of Si-O-Si[16].

Table 2. Chemical bonding on SiO2 from Mount Sinabung volcanic ash

No	Bands in SiO_2	Wavenumbers (cm^{-1})
1	O-Si-O (Bending)	470.63
2	Si-O (Bending)	798.53
3	Si-OH (Stretching)	948.98
4	Si-O-Si (Stretching)	1095.57
5	-OH (Bending)	1635.64
6	Si-C (Stretching)	2399.45
7	-OH (*Stretching*)	3437.15

3.4 XRD analysis

To find out the phase on SiO_2 of volcanic ash Mount Sinabung used XRD. The XRD measurement results can be seen in figure 3. SiO_2 has this amorphous properties can be seen in $2\theta = 22.16^o$. In addition to the formation of phases of SiO_2 formed another phase due to the coprecipitation process. The phase that appears is the phase of NaCl. This is in accordance with the XRF results that indicate the presence of C element in the SiO_2 content of Mount Sinabung volcanic ash.

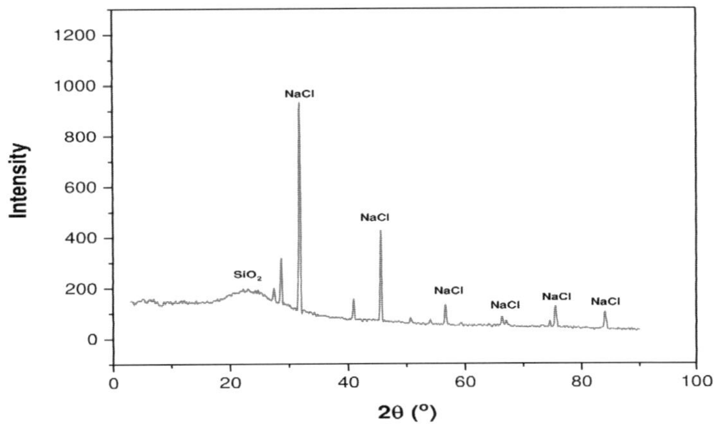

Figure 3. XRD image of SiO_2 from the Mount Sinabung volcanic ash.

4. Conclusions

Mount Sinabung Volcanic ash can produce Silicon dioxide (SiO_2) by coprecipitation method. SiO_2 obtained from the process as much as 48 %. From the results of SEM seen SiO_2 morphology that is still sequined with other minerals. Characterization using XRD obtained phases from SiO_2 which is amorphous. The characterization that indicates the existence of SiO_2 can be seen with the formation of Si-O-Si and Si-O function groups.

References

[1] D. O. Latif, A. Rifa'i, and K. B. Suryolelono, "Chemical characteristics of volcanic ash in indonesia for soil stabilization: Morphology and mineral content," Int. J. GEOMATE, vol. 11, no. 4, pp. 2606–2610, 2016, doi: 10.21660/2016.26.151120.

[2] L. Simatupang and D. S.Si, "The preparation and characterization of Sinabung volcanic ash as silica based adsorbent," J. Pendidik. Kim., vol. 8, no. 3, pp. 9–13, 2016, doi: 10.24114/jpkim.v8i3.4478.

[3] L. Simatupang, M. Situmorang, H. Marpaung, and R. Siburian, "Fabrication of silica-based chitosan biocomposite material from volcanic ash and shrimp husk by sol gel method for

adsorbent of cadmium (Ii) ions," Indian J. Chem. Technol., vol. 27, no. 5, pp. 387–394, 2020.
[4] R. Karolina, Syahrizal, M. A. Putra, and T. A. Prasetyo, "Optimization of the use of volcanic ash of Mount Sinabung eruption as the substitution for fine aggregate," Procedia Eng., vol. 125, pp. 669–674, 2015, doi: 10.1016/j.proeng.2015.11.102.
[5] H. Maulida, Ginting, M., dan Wici, "Menghasilkan Silika Gel Extraction Volcanic Ash of Sinabung Mount Silica To Production," J. Tek. Kim. USU, vol. 6, no. 3, pp. 41–46, 2017.
[6] S. Sulastri and S. Kristianingrum, "Berbagai Macam Senyawa Silika : Sintesis, Karakterisasi dan Pemanfaatan," Pros. Semin. Nas. Penelitian, Pendidik. dan Penerapan MIPA, pp. 211–216, 2010.
[7] L. Silvia and M. Zainuri, "Analisis Silika (SiO2) Hasil Kopresipitasi Berbasis Bahan Alam menggunakan Uji XRF dan XRD," J. Fis. dan Apl., vol. 16, no. 1, p. 12, 2020, doi: 10.12962/j24604682.v16i1.5322.
[8] I. Pratomo, S. Wardhani, and D. Purwonugroho, "Pengaruh Teknik Ekstraksi dan Konsentrasi HCl dalam Ekstraksi Silika dari Sekam Padi untuk Sintesis Silika Xerogel," Kim. Studentjournal, vol. 2, no. 1, pp. 358–364, 2013.
[9] H. Susilo, A. Putra, and A. Astuti, "Pengaruh Konsentrasi NaOH pada Sintesis Nanosilika dari Sinter Silika Mata Air Panas Sentral, Solok Selatan, Sumatera Barat dengan Metode Kopresipitasi," J. Fis. Unand, vol. 5, no. 4, pp. 334–338, 2016, doi: 10.25077/jfu.5.4.334-338.2016.
[10] P. Utari, N.P.S.N., Sudiarta, I.W., dan Suarya, "SINTESIS DAN KARAKTERISASI SILIKA GEL DARI ABU VULKANIK GUNUNG AGUNG MELALUI TEKNIK SOL-GEL N. P. S. N. Utari * , I W. Sudiarta, dan P. Suarya," J. Kim. (Journal Chem., vol. 14, no. 1, pp. 30–36, 2020.
[11] E. Dewa and R. Pasaribu, "Analisis kandungan Silikon dioksida (SiO2) pasir Pantai Koka Kabupaten Sikka dengan Metode Ekstraksi," Pros. Semin. Nas. Fis. PPs Univ. Negeri Makassar, vol. 2, pp. 76–79, 2020, [Online]. Available: https://ojs.unm.ac.id/semnasfisika/article/view/12857.
[12] N. Fuadi, "Analisis akndungan Silika (SiO2) Pada Butuan Green Tuff Dengan Metode Gravimetrik," Jambura Phys. J., vol. 2, no. 2, pp. 54–64, 2020.
1[3] S. Sembiring and P. Karo-Karo, "Pengaruh Suhu Sintering Terhadap Karakteristik Termal Dan Mikrostruktur Silika Sekam Padi," J. Sains MIPA., vol. 13, no. 3, pp. 233–239, 2007.
[14] S. N. Ishmah, M. D. Permana, M. L. Firdaus, and D. R. Eddy, "Extraction of Silica from Bengkulu Beach Sand using Alkali Fusion Method," PENDIPA J. Sci. Educ., vol. 4, no. 2, pp. 1–5, 2020, doi: 10.33369/pendipa.4.2.1-5.
[15] B. Shokri, M. A. Firouzjah, and S. I. Hosseini, "FTIR analysis of silicon dioxide thin film deposited by metal organic-based PECVD," Proc. 19th Int. Plasma Chem. Soc., no. December 2014, pp. 1–4, 2009, doi: www.ispc-conference.org.
[16] Masrur, Irmansyah, and Irzaman, "Optimasi Kelajuan Pemanasan Pada Ekstraksi," J. Biofisika, vol. 9, no. 2, pp. 13–20, 2013.

Acknowledgments

We would like thank you to Universitas Sumatera Utara and Universitas Asahan

IOP Conf. Series: Materials Science and Engineering 1156 (2021) 012016 doi:10.1088/1757-899X/1156/1/012016

Ecology and Disturbances in Bryophytes

Nurmaini Ginting[1,2*], Melvariani Syari Batubara[1], Muhammad Darwis[1], Vira Faradillah[3] and Melia Afnida Santi[3]

[1]Biology Education Study Program, Universitas Muhammadiyah Tapanuli Selatan, Indonesia.
[2]Universitas Sumatera Utara, Medan, Indonesia.
[3]Animal Husbandry Study Program, Universitas Muhammadiyah Tapanuli Selatan, Indonesia.

* nurmaini.ginting@um-tapsel.ac.id

Abstract. Bryophytes is a non vascular plant that has many benefits. However, the population and diversity of its kind experienced a significant decline. This decrease is caused by several factors such as deforestation, forest fragmentation, air and soil pollution and climate change. All these factors are caused by natural activities and mainly caused by human activities that make use of the environment without preserving nature.

1. Introduction

Indonesia is an archipelago on the equator. The land is covered by 120.6 million hectares of forest, or about 63% of the land area. Indonesia's location in the equator makes Indonesia has a tropical climate and becomes one of the countries with high biodiversity. In addition to its forest area, Indonesia is also covered by 14.9 million hectares of peatland[1].

Sumatra is one of the largest islands in Indonesia, 1,700 km long and 350 km wide. This area is dominated by Bukit Barisan mountains that stretch from the north end to the southern tip of Sumatra. The highest peak is Mount Kerinci with an altitude of 3,804 meters above sea level. Bukit barisan in the West is more humid when compared to the eastern part. Sumatra's diverse ecosystem makes Sumatra has a high diversity and endemicity. Some areas were later used as centersinsitu conservation areas in the form of National Parks [2]. However, currently the number of forest cover in Sumatra is decreasing, due to the transfer of forest functions for economic activities and human settlements [2].

Land use for economic activities poses a very big problem for forest areas, especially sumatran forests. The transition of forest functions into dwellings, highways, industrial estates and plantation land caused a drastic decline in forest area. Some people also do forest burning deliberately to open land, this forest fire became a very serious problem and became a concern of the international world. According to[2], Sumatra lost 3666.2 thousand hectares during the period 2009-2013, and became the area with the highest fire point. The most fires occurred in South Sumatra, then followed by North Sumatra and Riau. [3] also mentioned that forest fires in Indonesia greatly affect the flora and fauna that make the forest as their habitat. The Red Data List shows 772 species of endangered flora and fauna. One of the endangered flora is Bryophytes.

Deforestation activities greatly affect the survival of mosses especially epiphytic Bryophytes. Epiphytic Bryophytes depend on forest ecosystems, especially rainfall, temperature and humidity are greatly disrupted by land transfer, logging and forest fires. The morphology of Bryophytes and the way they breed are also strongly influenced by their habitat [3], [4]. This is also because, Bryophytes in the Asian region are more in the form of epiphytic Bryophytes attached to the bark and leaves of plants in the forest [5], [6].

2. Methods

This research uses literature study methods derived from journal articles that discuss about mosses and the factors that can influence their extinction. The articles taken mostly come from international publications, this is because moss research in Indonesia is still very rare.

3. Results and Discussion

There are many factors that affect the number of types of Bryophytes including deforestation, forest fragmentation, air pollution and climate change. Almost all of these factors are caused by humans. Some types of Bryophytes experienced a drastic decrease in number, resulting in almost extinction. [7] found that the *Sprutheanthustheobromae* (Lejeuneaceae) which is an endemic liverworth in cocoa plantations in the Dominican Republic is almost extinct. The reduction in moss population is due to the lack of care on the plantation.

The following will be presented several factors that affect the number of species and populations of Bryophytes in nature.

3.1 Deforestation

Deforestation is one of the most influential factors in biodiversity. In the tropics, many of its regions are converted into agricultural land. [8] conducted research in Sulawesi. He found that there is a difference in the number of types of moss in cocoa plantations and surrounding forests. This cocoa plantation comes from a forest that was cut down and converted into a garden. Drought-tolerant types of moss will be found in both locations. While sensitive types of moss are found only in forest areas [9] compared epiphytic Bryophytes in several habitats in Neotropic countries and Indonesia (Sulawesi), including rainforest areas, secondary forests, grasslands and cocoa plantations. They found there was a specific diversity of species between natural forest areas and disturbed forests. [10] added that drought-intolerant species have a profound effect on deforestation.

3.2 Fragmentation

Deforestation causes forest fragmentation in many forest areas in the tropics [11]. [12] found that there is an influence of forest fragmentation on ephipilous Bryophytes located in the Amazon. he stated there would be a reduction in the number of types in fragmented areas. In the wider fragment area will be found more varied types and high density of types. As Söderström research [13] found, habitat fragmentation is one of the problems that greatly affects species diversity in forests. Fragments with a large area will only reduce the number of species, but in the wider fragments will lead to the extinction of the species.

3.3 Air and Soil Pollution

In temperate climates, mosses are able to accumulate heavy metals [14], although in Bryophytes in the tropics research on this is still lacking [11]. [15], conducted research in Cezarina, Brazil. They found the presence of sulfur accumulation in Bryophytes tissues growing in areas with high sulfur content. These types fall into the group of Fabroniaceae, Hypnaceae and Helicophyllaceae.

Resarch on the accumulation of other heavy metals was conducted by [16] in the Srilanka mountain rainforest. The observed plants came from various groups, such as *Usnea barbata* (Lichen), *Pogonatumurnigerum* (Moss), *Lycopodium ulago* (fern), *Polypodium laneolatum* (Fern), *Bulbophyllumelliae* (Orchid), and *Actinodaphneambigua* (Spermatophyta). Of all the plants observed, elements as, Cd, Co and Pb are most commonly found in the moss tissue *Pogonatumurnigerum* (Figure 1).

Figure 1. Pogonatumurnigerum [11].

3.4 Climate Change

Climate change in the tropics has a profound effect on moss populations [11]. This happens because mosses are very sensitive to temperature. [17] research mentioned that a 1,5°C rise in temperatures in mountain forests in Bolivia could affect epiphytic moss communities within 2 years. So, mosses can help to see how quickly climate change is occurrence.

4. Conclusions

Bryophytes is a non vascular plant that has many benefits. However, the population and diversity of its kind experienced a significant decline. This decrease is caused by several factors such as deforestation, forest fragmentation, air and soil pollution and climate change. All these factors are caused by natural activities and mainly caused by human activities that make use of the environment without preserving nature.

References

[1] F. W. Indonesia, "Potret keadaan hutan indonesia," 2013.
[2] F. W. Indonesia, "Lembar Informasi Hutan Indonesia Yang teru s tergerus," no. November, 2014.
[3] N. Hallingback, T. hodgetts, *Status Survey and Conservation Action Plan for Bryophytes; Mosses, Liverworts, and Hornworts*. IUCN, Gland, Switzerland and Cambridge, UK in collaboration with the Swedish Threatened Species Unit, 2000.
[4] B. and benitez angel Maria, "Distribución potencial y áreas prioritarias para la conservación de briófitos en Ecuador Potential distribution and priority areas for conservation of bryophytes in Ecuador," vol. 10, no. 1, pp. 1–13, 2020.
[5] J. M. Glime and S. R. Gradstein, "CHAPTER 8-1 TROPICS : GENERAL ECOLOGY CHAPTER 8-1 TROPICS : GENERAL ECOLOGY," vol. 4, no. October, pp. 1–48, 2019.
[6] L. Soderstrom, T. Hallingbiick, and L. Gustafsson, "Bryophyte conservation for the future," vol. 3207, no. January 2016, 1992.
[7] S. R. Gradstein, S. P. Churchill, and N. Salazar, "Guide to the bryophytes of tropical America," *Mem. N. Y. Bot. Gard.*, vol. 86, no. January 2001, pp. 1–577, 2001.
[8] N. S. Ariyanti, M. M. Bos, K. Kartawinata, S. S. Tjitrosoedirdjo, E. Guhardja, and S. R. Gradstein, "Bryophytes on tree trunks in natural forests , selectively logged forests and cacao agroforests in Central Sulawesi , Indonesia," vol. 1, pp. 2–13, 2008.
[9] S. R. Gradstein and S. G. Sporn, "Land-use change and epiphytic bryophyte diversity in the Tropics Land-use change and epiphytic bryophyte diversity in the," no. April 2010, 2010.
[10] R. Gradstein and H. Culmsee, "Bryophyte diversity on tree trunks in montane forests of Central Sulawesi , Indonesia," no. November, 2009.

[11] J. M. Glime, "Chapter 8-14 Tropics : Disturbance And Conservation Chapter 8-14 Tropics : Disturbance," vol. 4, no. October, pp. 1–34, 2019.
[12] C. E. Zartman, "Habitat fragmentation impacts on epiphyllous Bryophyte communities in Central Amazonia," vol. 9658, no. January, 2019.
[13] L. Söderström, "Conservation biology of bryophytes Conservation biology of Bryophytes," no. May, 2014.
[14] T. Faus-kessler, C. Dietl, J. Tritschler, and L. Peichl, "Correlation patterns of metals in the epiphytic moss Hypnum cupressiforme in Bavaria," vol. 35, pp. 427–439, 2001.
[15] A. A. Rodrigues, D. A. Rodrigues, and O. Moura, "Bryophytes in areas subject to exposure to sulfur dioxide Bryophytes in Areas Subject to Exposure to Sulfur Dioxide," no. October, 2015.
[16] R. Jayasekera and M. Rossbach, "Background levels of heavy metals in plants of different taxonomic groups from a montane rain forest in Sri Lanka," no. October 2016, 1996.
[17] J. Jacome, S. R. Gradstein, and M. Kessler, "Responses of epiphytic bryophyte communities to simulated climate change in the Tropics Responses of Epiphytic Bryophyte Communities to Simulated Climate Change in the Tropics," no. January, 2011.

IOP Conf. Series: Materials Science and Engineering 1156 (2021) 012017 doi:10.1088/1757-899X/1156/1/012017

Response of Oil Palm Plant Growth (*Elaeis guineensis Jacq.*) to Aluminium Stress

Nurul Huda Panggabean[1]*, Isnaini Nurwahyuni[2], Elimasni Djamaan[2]

[1]STKIP Asy – Syafi'iyah Internasional Medan,
[2]University of Sumatera Utara

*nurulhudapanggabean@gmail.com

Abstract. This research aims to study the response of palm oil plants to Aluminum stress their deficiencies. This experiment used Aluminum (Al) with different concentrations of 0 ppm (Al0), 100 ppm (Al1) and 200 ppm (Al2) dissolved in hoagland liquid media with different old of oil palm plants. Based on the results obtained, the effect of aluminum stress on the wet weight of the plant has a real effect while it has the same effect on plant height and plant root length. Some deficiencies experienced by palm oil plants that are in stress of Al, the first symptom that is seen is the plant roots become shorter and thicker, as a result the growth and development of roots is inhibited and in the long term can cause the ability of plants to absorb nutrients is reduced. Anatomically observations of chlorophyll content and stomata density, where plants aged 6 months have the highest total chlorophyll content (0.059602, 0.05612, and 0.066077) while the lowest chlorophyll content in plants is 2 months old (0.018388, 0 , 022428 and 0.017978). The density of stomata at 2 months age has the highest values while the lowest density values for plants in 6 months for each treatment

1. Introduction

Palm oil (Elaeis guineensis Jacq.) is the most productive vegetable oil producing plant in the world when compared to other types of vegetable oil producing plants such as canola, coconut, soybean and sunflower. Palm oil is the main crop of plantations in Indonesia, is a plantation plant which occupies an important position in the agricultural sector in general and the plantation sector in particular. This can be seen from the increase in plantation area, oil production and commodity export volume from year to year. Indonesia's palm oil production centers mainly from 6 (six) provinces which contributed 73.69% of the total Indonesian palm oil production. The provinces of Riau and North Sumatra are the largest CPO production centers in Indonesia with contributions of 23.75% and 16.24%, respectively. [1]Palm oil is an annual plant that can reproduce economically until the age of 25-30 years, to support this potential maintenance of vegetative growth is important. Potential land for oil palm cultivation is limited by environmental factors, physical and chemical properties of the soil. Soil fertility is a condition of the ability of the soil to provide adequate and balanced amounts of nutrients so that it supports plant growth and production according to its potential [2]More than 55 million hectares of agricultural land in Indonesia is acidic. Aluminum (Al) is known as a major factor causing toxicity for plants that grow in acidic soils. Vulnerability of plants to Al stress will cause plants to be susceptible to drought and disruption of nutrient uptake, so that in the long run it will affect their growth and development. Aluminum is one of the supporting nutrients that can cause soil poisoning around plant roots, so that it can inhibit plant growth and development [3]According to [4] Al causes disruption of cell division in the root cap and lateral roots and causes an increase in cell rigidity through the formation of pectin cross bonds in the cell wall, reducing DNA replication through increased double chain rigidity.

Aluminum is a rhizotoxic ion that inhibits plant growth and productivity in acidic mineral soils. Although Al inhibits the process of metabolism and plant growth, but to some extent the influence of Al can be tolerated by tolerant plants [5].The initial symptom that appears in plants that are Al poisoning is a root system that is not well developed [6].Several studies on the effect of Al on plant root systems have been widely reported, but the mechanism of growth inhibition induced by Al is not widely understood. Plasma membrane of cells at the root tip is the main target of Al toxicity [7]The main influence of Al on root membrane permeability can be seen only within minutes or hours after Al exposure.

2. Research Methods

2.1 Time and Place of Research

This research was conducted in June 2020 at the Laboratory of Physiology and Plant Tissue Culture, Department of Biology, Faculty of Mathematics and Natural Sciences, University of North Sumatra

2.2 Plant Preparation and Growing Media

The plants used are palm oil (Elaeis guineensis Jacq.) with varieties of D X P Langkat which are 2 months, 4 months and 6 months old. Plants come from sprouts planted on sand and soil media in the pre nursery stage. Plants are then transferred from the soil growing media to the hoagland liquid media that was previously cleaned from dirt attached to running water. Aeration is done every day in 2 weeks. The form of Al salt used as a treatment is Al-nitrate salt. The seeds were monitored for growth for 2 weeks looking at plant deficiencies and responses.

2.3 Treatment

The treatments in this experiment used Aluminum (Al) with different concentrations of 0 ppm (Al0), 100 ppm (Al1) and 200 ppm (Al2) dissolved in hoagland liquid media with 5 replications per treatment for one type of plant life of 2 months (U1), 4 months (U2) and 6 months (U3).

Table 1. Design of Aluminum (Al) Concentration Treatment Treatments Plant age (U)

Umur Tanaman	**Ulangan**	**Konsentrasi Aluminium (Al)**		
		Al_0	**Al_1**	**Al_2**
	1	Al_0U_1	Al_1U_1	Al_2U_1
	2	Al_0U_1	Al_1U_1	Al_2U_1
2 bulan	3	Al_0U_1	Al_1U_1	Al_2U_1
	4	Al_0U_1	Al_1U_1	Al_2U_1
	5	Al_0U_1	Al_1U_1	Al_2U_1
	1	Al_0U_2	Al_1U_2	Al_2U_2
	2	Al_0U_2	Al_1U_2	Al_2U_2
4 bulan	3	Al_0U_2	Al_1U_2	Al_2U_2
	4	Al_0U_2	Al_1U_2	Al_2U_2
	5	Al_0U_2	Al_1U_2	Al_2U_2
	1	Al_0U_3	Al_1U_3	Al_2U_3
	2	Al_0U_3	Al_1U_3	Al_2U_3
6 bulan	3	Al_0U_3	Al_1U_3	Al_2U_3
	4	Al_0U_3	Al_1U_3	Al_2U_3
	5	Al_0U_3	Al_1U_3	Al_2U_3

2.4 *Observation Parameters*
The parameters in this experiment include measurements of plant height, root length, final wet weight and observation of stomata and chlorophyll, as well as plant morphology after treatment.

2.5 *Analysis of stomata density and leaf chlorophyll*
The stomata density of each leaf sample was calculated based on the method of [8]i.e. by applying a nail polish solution on the underside of the leaf. The dried nail polish is removed with the aid of masking tape and then glued to a glass preparation which has been dripped with iodine solution as a coloring agent. The number of stomata was calculated using a microscope at 400x magnification.

Chlorophyll content was measured using spectrophotometric methods. The leaves used are crushed using a mortar, weighed 0.05 grams. Samples were extracted with 5 ml of 70% acetone stirred until chlorophyll dissolved. Chlorophyll extract was filtered using filter paper. The filtrate obtained was added to the cuvette to further measure the chlorophyll content. The instrument used is a spectrophotometer at wavelengths of 645 nm and 663 nm, after absorbing values, chlorophyll a, chlorophyll b and total chlorophyll are calculated.

3. Results and Discussion

3.1 *Wet Plant Weight Parameters*
The results of analysis of variance for the parameters of wet weight, plant height and root length showed different results for each parameter observed. Aluminum stress and age significantly affect the wet weight of oil palm plants. This can be seen from the response of oil palm plants which are in stress conditions Al1 and Al2 which have lower wet weight (5.7460 and 5.7193) compared to plants that are in conditions without Al stress which is 7.0320.

Wet weight is a parameter that shows the accumulation of photosynthate and water in the body of the plant. Wet weight is calculated by weighing the plant part immediately after it is removed from the growing media. This is because water is so mobile that it can move from one part to another or is lost through transpiration produced by plants [9]

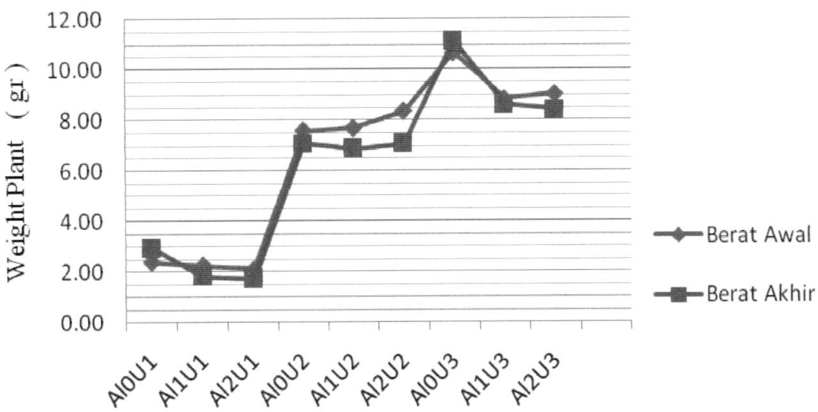

Figure 1. Wet Weight Of Oil Palm

From the graph above, there can be a decrease in plant wet weight in aluminum stress, this is caused by damage to the plant root system that is in Al so that the ability of plants to absorb water is reduced. Plant fresh weight can indicate plant metabolic activity and plant fresh weight value is influenced by tissue water content, nutrients and metabolic yield. Decreasing the size and number of cells will also ultimately reduce the weight of plants [10].

3.2 Plant Height Parameters

Plant height is a plant growth variable that is easily observed as a parameter to determine the effect of the environment or the effect of treatment on plants. Based on observations From each treatment the same effect on plant height at all ages. However, there was a decrease in plant height in Al2 treatment when compared to plants in Al0 and Al1 which were not much different. This shows the intolerance of plants to Al concentrations of more than 100 ppm.

Plant height is the result of cell division and elongation. Plant growth is the result of various physiological processes that involve genotypic factors that interact with environmental factors. Growth occurs in terms of increasing size, shape and amount. Growth on plant stems is due to the activity of the apical meristem located at the tip of the shoots. Meristems are composed of initial cells that play the role of forming new cells and derivative cells that will be specialized into certain tissues [2].

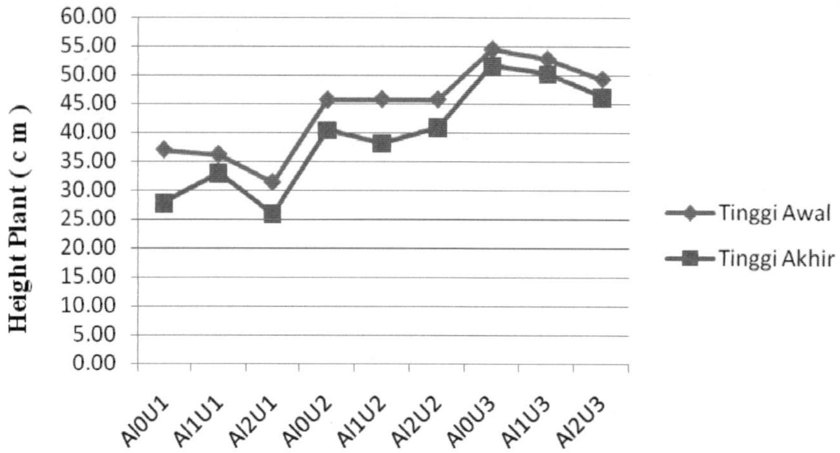

Figure 2. Height Of Oil Palm Plants

3.3 Plant Root Length Parameters

the influence of aluminum stress and age have the same effect on the root length of the plant, but when viewed from the average root length there is still a decrease in root length that does not differ greatly from each other between each treatment. This can be seen in Figure 3.

Plants that are sensitive to Al experience inhibition of plant growth and development, due to stunted root growth. The first symptom that appears from Al poisoning is the root system that does not develop (short and thick) due to inhibition of cell extension [11].

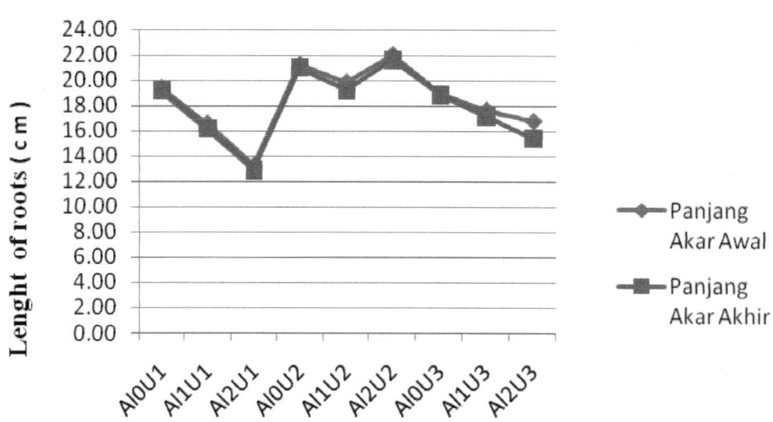

Figure 3. Length of Oil Palm Roots

Inhibition of plant growth due to Al stress due to the formation of bonds between Al with the root plasma membrane and on the root cell wall and can replace the Ca position in the middle lamella all of which cause inhibition of cell division and root function. Inhibition of root formation in plants that have stressed Al causes a decrease in root length and cell viability because 99% of Al that accumulates in cells is found in cell walls and membranes, related to compounds such as phospholipids found in cell membranes that interfere with membrane permeability and interfere with absorption nutrients regulated by a proton pump. Damage to the roots results in inhibition of absorption of nutrients and water in the soil. Imperfect root growth causes the root system to become shallower and more sensitive to drought. High enough aluminum concentrations in the planting medium can inhibit the growth of some species, not only because their effects damage the availability of phosphate, it also seems to inhibit iron absorption and because of their direct toxic effects on plant metabolism. Al poisoning inhibits the extension and growth of primary roots, as well as blocking the formation of lateral roots and root hairs [12]

3.4 Response of Physiology Variables

3.4.1 Chlorophyll content

From the results obtained in this experiment it can be seen that oil palm plants aged 6 months for each treatment (Control, Al 100 ppm and Al 200 ppm) have the highest total chlorophyll content (0.059602, 0.05612, and 0.066077) while the lowest chlorophyll content in plants aged 2 months for each treatment is (0.018388, 0.022428 and 0.017978) so it can be concluded that plant age affects the chlorophyll content of the plant.

With the presence of Al stress in the planting medium the ability of the root to absorb Ca2 is replaced by competitor Al so that root growth is inhibited and this has implications for low Mg2 + uptake and inhibits the formation of chlorophyll [2] The decrease in chlorophyll content that occurs due to damage to roots causes inhibition of absorption of nutrients and water in the soil. Imperfect root growth causes the root system to become shallower and more sensitive to drought. so that if a plant suffers from water shortages due to damage to the root system it will inhibit chlorophyll synthesis in leaves, due to decreased photosynthesis rate and an increase in temperature and respiratory can cause chlorophyll disintegration [13].

3.4.2 Stomata Density

Observation of stomata density in oil palm plants showed that the age of 2 months for each treatment had the highest value (649.90, 524.11, and 507.34) while the lowest density value for 6 months for each treatment (557.65, 511, 53 and 473.79). Stomata are the most important part in the exchange of gases (O2 and CO2) in plants, which is one of the epidermal modifications that are in the abaxial leaves. Stomata density in oil palm plants can be categorized as high density (> 500 / mm2) only in plants aged 6 months with stress Al 200 ppm categorized as medium density (300-500 / mm2).

Stomata develop gradually during organ growth, so that younger organs have fewer stomata than adult organs. Stomata density is closely related to the metabolic or physiological processes of plants (Khoiroh.2014). from the category of stomata density which tends to be high in oil palm plants with stress Al shows a physiological response related to the inhibition of water absorption by roots, where plants that grow in areas that lack water will have a greater stomata density compared to plants that live in the area wet and protected. Plants that have a large stomata density will have a higher transpiration rate than plants with a low stomata density [14].

4. Conclusions

Aluminum stress on oil palm (Elaeis guineensis Jacq.) Aged 2 months, 4 months and 6 months within 10 HST in Hoagland media significantly affected the wet weight of plants and had the same effect on root height and length. There is a decrease from observations of plant height, root length and wet weight of plants due to plant response to Al. Al deficiency can be seen from the morphological changes in plant roots which become shorter, thicker and stiffer. Physiologically the roots experience growth and development that is inhibited resulting in the absorption of nutrients for plants is also reduced. In addition, due to Al poisoning by plants, plants are also vulnerable to drought stress. Experiments need to be carried out for oil palm plants with older plant age and the determination of the appropriate liquid media in determining the effect of Al stress on plants.

References

[1] Pusat Data dan Sistem Informasi Pertanian, " Outlook Kelapa Sawit.," Kementrian Pertanian, 2016.

[2] L. S. Mardhika, "Respon Pertumbuhan Tanaman Kelapa Sawit (Elaeis guineensis Jacq.) Belum Menghasilkan Umur Dua Tahun Terhadap Pemupukan Kalsium," Bul. Agrohorti , vol. 3, no. 1, pp. 110 - 118, 2015.

[3] S. Anwar, " Toleransi Morfologi Dan Fisiologi Tanaman Rumput Pakan Terhadap Cekaman Aluminium.," J. Indon.Trop.Anim.Agric, vol. 28(1), 2003.

[4] Silva, I. R., Smyth, T. J., Moxley, D.F., Carter, T.E., Allen, N.S. & Rufty, T.W, "[13] Aluminum accumulation at nuclei of cells in the root tip. Fluorescence detection using lumogallion and confocal laser scanning microscopy," Plant Physio, vol. 123, p. 543, 2000.

[5] M. Utama, "Mekanisme Fisiologi Toleransi Cekaman Aluminium Spesies Legum Penutup Tanah Terhadap Metabolisme Nitrat (NO3-), Ammonium (NH4) dan nitrit (NO2-)," Bul. Agron, vol. (36) (2) , p. 176 – 180, 2008.

[6] Zhang, X., Liu, P., Yang, Y. S. & Xu, G. D., "[16 Effect of Al in Soil on Photosynthesis and related morphological and physiological characteristics of two soybean genotypes," Bot. Stud, vol. 48, p. 435–444, 2007.

[7] Takabatake, & Shimmen, R.T, "Inhibition of electrogenesis by aluminum in characean cells.," Plant Cell Physiol, vol. 38, pp. 1264-1271, 1997.

[8] Pangaribuan Y, Sudradjat, Asmono D., "Respon Fisiologi Beberapa Varietas Kelapa Sawit di Pembibitan Terhadap Cekaman Air," Jurnal Penelitian Kelapa Sawit, vol. 8(2), p. 81 – 95, 2000.

[9] S. Tuhuteru, " Kajian Fisiologis Tanaman Tomat Terhadap Penambahan Unsur Hara Fe dan N," Jurnal Agroekotek , vol. 10, no. 2, pp. 64 - 72, 2018.

[10] V. Fatimah, "Respon Karakter Fisiologis Kedelai (Glycine max L.) Varietas Grobogan Terhadap Cekaman Genangan," Institut Teknologi Sepuluh November, Surabaya, 2016.

[11] Pujiwati, H. Ghulamahdi Munif. Yahya Sudirman. Aziz Arifin Sandra. Haridjaja Oteng, "Tanggap Kedelai Hitam Terhadap Cekaman Alumunium Pada Kultur Hara," Penelitian Pertanian Tanaman Pangan, vol. 35, no. 2, 2016.

[12] D. Sofia, " Respon Tanaman Kedelai (Glycine max (L.) Merril) Pada Tanah Masam.," Fakultas Pertanian, Universitas Sumatera Utara, Medan, 2007.

[13] E. T. S. M. A. I. Febrianto, " Evaluasi Karakter Agronomi Kelapa Sawit (Elaeis guineensis Jacq.) Varietas D x P Avros Pada Kondisi Cekaman Kekeringan Di Main Nursery," BERNAS Agricultural Research Journal, vol. 15 No. 1, 2019.

[14] L. Juairiah, "Studi Karakteristik Stomata Beberapa Jenis Tanaman Revegetasi Di Lahan Pasca Penambangan Timah Di Bangka," Widyariset, vol. 17 No.2, 2014.

Water Quality Parameters of Tofu Wastewater: A Review

Primadi Gayuh Laksono Putro*, **H. Hadiyanto and Amirudin**

Graduate Program, Universitas Diponegoro, Semarang, Indonesia

* primadigayuh@gmail.com

Abstract. The tofu industry's existence is quite potential in recruiting labors, which can improve the economy of the surrounding community. On the other hand, it can also have a negative impact due to the wastewater produced from the tofu making process, which can damage the environment. Water is widely used as material for washing and boiling soybeans; due to the large use of water in making tofu and tempeh, fermented soybeans. The resulting waste is also quite large. This research presents a literature review with primary data on articles, magazines, reports, and theories. The purpose of this research is to provide an overview of the water quality parameters of tofu waste. The liquid waste produced by the tofu-making industry is a thick liquid that is separated from the clumps of tofu, which is called whey. This liquid contains high levels of protein and can quickly break down. This waste is often disposed of directly without prior treatment, resulting in a bad smell and polluting the environment. There are several parameters used to measure water quality, including physical, chemical, and biological properties.

1. Introduction

Tofu is one of the most popular foods among Indonesians. This food is favored by almost all levels of society, lower, middle, and upper classes. It can be seen from the sale of tofu in all conditions, vegetable peddlers, markets, supermarkets, restaurants, and star hotels. One of the factors why it is liked by the public is because it is affordable and delicious. Tofu also has good nutritional contents such as protein and carbohydrates. In general, one is used as a side dish and snack.

The problem that arises in SME is the abundance of solid and liquid waste generated during production. Those wastes are still not properly managed, such as solid waste that can be reprocessed into soft tempeh or *gembus*, flour, and cakes. The water disposal system should be separated between liquid waste generated by the production installation and household liquid waste. This separation aims to classify waste management methods according to the source of the pollution. The liquid waste will react with the air and decompose by microorganisms, causing problems in the form of an unpleasant odor that pollutes the surrounding environment and water.

According to [1], if every quintal of soybeans used for making tofu produces 1.5 - 3 m^3 of wastewater. For example, the amount of soybeans with a value of 116,290 kg = 1162.9 quintals produces wastewater ranging from 1162.9 quintals x 2 m^3 = 2325.8 m^3 / day = 69,774 m^3 / month = 837,288 m^3 / year of wastewater produced. Therefore, it becomes a serious problem if the large volume of waste produced exceeds the carrying capacity of the environment. The definition of the carrying capacity of the environment according to [2] is the amount of the ability of the environment to support the life of organisms expressed in space units.

In this paper, the researcher discusses how the impacts of tofu waste on the environment and the parameters for measuring the water quality of tofu waste. The purpose of this article is to provide a broader view of the water quality parameters of tofu waste.

2. Method

This research is a review, descriptive, and based on the non-empirical design [3], [4]. The collected information of scope was conducted in secondary sources of books, articles, journals, electronic sources, and theories. A review paper provides a summary and synthesis of research findings on a topic selected and completed by other authors [5]. The research review aims to find out the current state of relevant publications regarding the findings related to the chosen topic. This research is a descriptive study, an analysis that describes a phenomenon and its characteristics systematically, factually, and accurately [6]. One uses non-empirical design or also called theoretical research [7][8], which is not based on empirical evidence such as quantitative data or qualitative data but is still based on facts [9]. In this research, the researchers chose the topic of tofu wastewater and water quality, so the researchers looked for various articles that corresponded to the issue of water quality parameters of tofu wastewater from publications that have been published by other authors.

3. Discussion

3.1 Tofu Waste

Tofu waste comes from waste or residual processing of soybeans which is wasted because it is not properly solid so it cannot be consumed. It consists of two types, namely liquid and solid waste. This waste occurs because of the remaining tofu water that does not clot, pieces of tofu that are destroyed due to the imperfect clumping process, and yellowish liquid which can cause an unpleasant odor [10].

The solid waste of the tofu processing plant is in the form of dirt from cleaning soybeans (stones, soil, soybean skin, and other solid objects that stick to the soybeans) and the residue from the soybean pulps filter called tofu pulp. The solid waste in the form of manure comes from the initial process (washing) of soybean raw materials and generally, there is not so much solid waste (0.3% of the raw material for soybeans). Meanwhile, solid waste in the form of tofu dregs occurs in the soybean pulp filtering process. The number of tofu dregs that is formed ranges from 25-35% of the tofu produced [11].This liquid contains high levels of protein and can break down quickly. This waste is often disposed of directly without prior treatment, resulting in a bad smell and polluting the environment [11].

3.2 Water Quality

According to [12], the parameters used to measure water quality include physical, chemical, and biological properties. These parameters are:

3.2.1 Physical Properties

The physical parameters of water that determine its quality are turbidity, temperature, color, smell, taste, amount of suspended solids, dissolved solids, and electrical conductivity (EC).

3.2.2 Chemical Properties

Chemical properties that can be used as indicators that determine water quality are pH, concentrations of potassium, magnesium, manganese, iron, sulfide, sulfate, ammonia, nitrite, nitrate, phosphate, dissolved oxygen, BOD, COD, oil, fat, and heavy metal.

3.2.3 Biological (Microbiological) Properties

Organisms in water can be used as indicators of pollution in an aquatic environment, for example bacteria, algae, plankton, and certain fish. The method of measurement taken for each parameter varies according to the circumstances [13]–[15].

1. Water Temperature, The temperature of the wastewater produced usually has a higher temperature than water in public drains[13]. DO in hot water is relatively small, so it also decreases in the common sewer where the tofu wastewater is disposed.

2. pH, is divided into acidic and alkaline properties[16]. Acidity is the ability to neutralize bases. Weak acidity can have high acidity. It means that it has the potential to release hydrogen. It is differentiated between free acidity and total acidity. Free acidity is caused by strong acids such as hydrochloric acid and sulfuric. Free acidity can lower pH. Total acidity consists of the free acidity and the acidity caused by weak acids. Meanwhile, the basicity of water is the capacity of water to neutralize acids. This is because there are bases or alkaline salts in the water. For example, NaOH, Ca $(OH)_2$, and so on. Basic salts that are often encountered are carbonates of the metals sodium, calcium, magnesium, and so on. While high acidity does not mean high pH.To fulfill the requirements of life, water must have a pH of about 6.5 - 7.5. It will be acidic or base depending on the big or small pH. If the pH <6.5 then the water is acidic, while water that has a pH$>$ 7.5 is alkaline or base. Wastewater and industrial waste will change the pH of the water which will ultimately disrupt the life of aquatic biota which is sensitive to changes in pH [17].
3. TSS, Total solids suspended in water are inorganic, organic, and liquid particles that cannot mix in water. It includes soil, clay, and mud, while organic solid often encountered are plant fibers, algae cells, and bacteria. These solids are natural pollutants originating from water erosion while flowing [18].Other suspended residue compounds come from the activities of the population using water. Residual waste and industrial waste usually contain a lot of suspended residues. The presence of suspended residues in water is undesirable for reasons of decreasing aesthetics of water, besides the suspended residues can be used as absorption sites for chemical or biological substances such as disease-causing microorganisms [19].The maximum limit of suspended solids content in water for Class I water body designation is 50 mg/ liter (PPRI-82, 2001).
4. BOD, BOD is the amount of oxygen needed by microorganisms to degrade the organic material in water [17]. According to [20], one is the amount of oxygen expressed in mg / l or parts per million (ppm) used by bacteria to oxidize in water. Organic material consisting of carbohydrates (cellulose, starch, sugar), protein, hydrocarbon oil, and other organic materials enters the water body from natural sources as well as from polluting sources. The natural sources of BOD in surface water come from plant rot and animal waste, while the source of BOD from human activities comes from feces, urine, detergents, oils, and fats [21]. In general, BOD is used to determine the level of wastewater pollution. Its measurement is a measurement of the amount of oxygen used by microorganisms in decomposing organic matter in water. The oxidation reaction during the BOD examination is the result of biological activity and the reaction that takes place is influenced by the number of populations and temperature. The temperature should be kept constant at 20°C which is the common temperature in nature. Theoretically, the time required for the complete oxidation process to break down organic material into CO_2 and H_2O is unlimited. In laboratory practice, it usually lasts for 5 days with the assumption that during that time the percentage of reactions is quite large from the total BOD [22]. If the high O_2 consumption is indicated by the smaller dissolved O_2, it means that the content of waste materials requires high O_2[23].
5. COD, COD is defined as the amount of oxygen required for the waste materials present in water to be chemically oxidized. The organic waste material will be oxidized by Potassium Bichromate to CO_2 and H_2O gas to form Chrom ions. Potassium Bichromate is used as a source of oxygen (oxidizing agent) following the reaction: 22 CaHbOc + Cr_2O72-□□CO_2 + H_2O + Cr_3 (Source: [17]). The amount of oxygen required for the oxidation reaction against organic waste is the same as the amount of potassium bichromate used in the oxidation reaction [17]. Waters that have a high COD value are undesirable for fisheries and agriculture purposes. COD values in uncontaminated waters are usually less than 20 mg / l [15]. Its maximum levels are allowed to support aquatic organisms and for irrigation purposes range from 10 - 100 mg / l (PP No. 82 of 2001 concerning Water Quality Management and Water Pollution Control).
6. DO, Dissolved oxygen is important to use to decompose or oxidize organic and inorganic materials in aerobic processes in water. The main source of oxygen in waters comes from the air through the diffusion process and the photosynthesis of organisms in these waters [22]. In freshwater, dissolved oxygen levels at 25°C are around 8 mg / l [15]. According to [16], in flowing waters, usually oxygen is not a limiting factor. In a river that is clear and swift, the oxygen

concentration reaches saturation. Dissolved oxygen determinations must be carried out multiple times, at various locations, at different depth levels at different times. If the water is running slowly or there are pollutants then the dissolved oxygen may be undersaturation, so the oxygen again becomes the limiting factor. It depends on (1) temperature; (2) the presence of photosynthetic plants; (3) The level of light penetration depends on the depth and turbidity of the water; (4) Water hardness level; (5) The amount of organic material that has been broken down in water such as garbage, dead algae, or industrial waste.

7. NH_3, In the waters, nitrogen is in the form of inorganic and organic. Inorganic nitrogen consists of ammonia (NH_3), ammonium (NH_4), nitrite (NO_2), nitrate (NO_3), and nitrogen molecules (N_2) in gaseous form. Organic nitrogen is in the form of protein, amino acids, and urea. The main source of anthropogenic nitrogen in waters comes from agricultural areas that use fertilizers intensively and from domestic activities. Ammonia and its salts are soluble in water. The source of ammonia in the waters is the breakdown of organic nitrogen (protein and urea) and inorganic nitrogen found in soil and water, which comes from the decomposition of dead plant and aquatic biota by microbes and fungi [15]. The free ammonia content for drinking water purposes should not be more than 0.5 mg / l, while for fisheries, the free ammonia content for sensitive fish is ≤ 0.02 mg / l as NH_3 (Government Regulation No. 82 of 2001 concerning Water Quality Management and Water Pollution Control).1) $2NH_4 + + 3O_2$ + Nitrosomonas $2NO_2$- + $2H_2O$ + 4H + + Energy2) $2NO_2 + O_2$ + Nitrobacter $2NO_3$ + Energy (Source: [15]).

The presence of nitrite content in the waste shows that little of the organic nitrogen compounds are oxidized. It is only a little in new waste, but in stale waste, a large amount of nitrite is found. The presence of nitrites indicates that change is taking place, thereby indicating imperfect waste management. Nitrate (NO_3) and ammonium (NH_4) are the main sources of nitrogen in the water, but NH4 is preferred by plants. One represents the end product of the degradation of organic material (nitrogen). Nitrate-nitrogen levels in natural waters are almost never more than 0.1 mg / l. NO_3 levels of more than 5 mg / l represent the occurrence of anthropogenic contamination originating from human activities and animal feces. Nitrate-nitrogen levels of more than 0.2 mg / l can result in eutrophication (enrichment) of water, which in turn stimulates the rapid growth of algae and aquatic plants (blooming). In waters that receive runoff from agricultural areas that contain lots of fertilizers, nitrate levels can reach 1000 mg / l. Nitrate levels for drinking water should not exceed 10 mg / l (PP 82 of 2001 concerning Water Quality Management and Water Pollution Control). Adults have a high tolerance for nitrate ions, but for babies and animals, these ions are toxic. In the digestive systems of babies and animals, nitrates are reduced by nitrites. Nitrite can bind to hemoglobin in the blood [24]. The N content in water as total nitrogen (N), dissolved nitrogen (DIN), nitrate (NO_3-N), and ammonium (NH_4-N) increases along with the rainy season. Rainfall and water runoff are the main drivers that cause N from non-point sources to be released from their catchment areas, while fertilizers cause the input of large amounts of N into the environment and agricultural activities accelerate the transformation of N into water bodies [25].

The river as intended must always be in its condition, namely by being protected and preserved; enhanced functions and benefits; and the destructive power of the environment is controlled [26].This condition will cause all dissolved pollutants in the form of liquid and solid waste to enter the river flow. The number of pollutants that enter the river will affect the quality of river water. At a certain point, there will be pollution [14].

3.3 Community Behavior

Behavior is the result of all human experiences and interactions with the environment which are manifested in the form of knowledge, attitudes, and actions[27]. One is formed through a process and takes place in human and environmental interactions. It is formed through internal and external factors. Internal factors include knowledge, intelligence, emotion, and innovation. Meanwhile, external factors include the surrounding environment, both physical and non-physical, such as climate, socio-economy, and culture. The behavior is a person's response or reaction to external stimulus[28]. Therefore it occurs through the process of a stimulus to the organism, then the organism responds, so the SKINNER theory is called an S-O-R theory or Stimulus Organism Response. Community behavior is the process of evaluating a person towards an object, event, or stimulus by involving experiences related to that object through processes of cognition and affection to form an object. Besides, behavior is influenced by three factors, namely predisposing, predisposing, enabling, and reinforcing factors[29]. Predisposing factors are positive factors that facilitate the practice. These factors include knowledge, attitudes, beliefs, traditions, social norms, experience, education level, beliefs, and other forms that exist in individuals/communities.

4. Conclusion

Tofu industrial waste is generally divided into two forms of waste, namely solid and liquid waste. There are three parameters that can be used to measure water quality, namely physical, chemical, and biological properties. Physical properties include turbidity, temperature, color, smell, taste, amount of suspended solids, dissolved solids, and electrical conductivity (EC). Chemical properties focus on the pH of potassium, magnesium, manganese, iron, sulfide, sulfate, ammonia, nitrite, nitrate, phosphate, dissolved oxygen, BOD, COD, oil, fat, and heavy metal. Meanwhile, the biological properties (microbiology) include water temperature, pH, TSS, BOD, COD, DO, and NH_3. Finally, the community behavior pattern in the tofu production process also needs attention. One is formed through internal and external factors. The internal factors include knowledge, intelligence, emotions, and innovation. Meanwhile, the external factors include the surrounding environment, both physical and non-physical, such as climate, socio-economy, and culture.

5. References

[1] P. B. Aji and J. Kimia, "PENURUNAN NILAI COD AIR LIMBAH PABRIK TAHU Info Artikel," *Indones. J. Chem. Sci.*, vol. 5, no. 2, pp. 2–5, 2016.

[2] O. Soemarwoto, *Ekologi, Lingkungan Hidup Dan Pembangunan*. Jakarta: Djambatan, 2004.

[3] S. Apriyanto, Dalman, and D. Santoso, "The urgency of forensic linguistics in a police interrogation process," *Int. J. Psychosoc. Rehabil.*, vol. 24, no. 6, pp. 4766–4772, 2020, doi: 10.37200/IJPR/V24I6/PR260467.

[4] O. Hidayat, S. Apriyanto, P. Program, A. Science, U. Tun, and H. Onn, "Drama Excerpt : Tool in Enhancing Speaking Ability for Junior High School," vol. 2, no. 3, pp. 1–9, 2019.

[5] R. W. Palmatier, M. B. Houston, and J. Hulland, "Review articles: purpose, process, and structure," *J. Acad. Mark. Sci.*, vol. 46, no. 1, pp. 1–5, 2018, doi: 10.1007/s11747-017-0563-4.

[6] H. Nassaji, "Qualitative and descriptive research: Data type versus data analysis," *Lang. Teach. Res.*, vol. 19, no. 2, pp. 129–132, 2015, doi: 10.1177/1362168815572747.

[7] R. C. GARDNER and P. F. TREMBLAY, "On Motivation, Research Agendas, and Theoretical Frameworks," *Mod. Lang. J.*, vol. 78, no. 3, pp. 359–368, 1994, doi: 10.1111/j.1540-4781.1994.tb02050.x.

[8] K. Moen and A. L. Middelthon, "Qualitative Research Methods," in *Research in Medical and Biological Sciences: From Planning and Preparation to Grant Application and Publication*, Australia: Elsevier Ltd, 2015, pp. 321–378.

[9] O. O. Fidelis, "Writing non-empirical articles for publication," *Int. J. Adv. Acad. Res.*, vol. 3, no. 3, pp. 25–31, 2017.

[10] Nohong, "Pemanfaatan Limbah Tahu sebagai Bahan Penyerap Logam Krom , Kadmiun dan Besi Dalam Air Lindi TPA," *J. Pembelajaran Sains*, vol. 6, no. 2, pp. 257–269, 2010.

[11] F. Kaswinarni, "Kajian Teknis Pengolahan Limbah Padat Dan Cair Industri Tahu (Studi Kasus Industri Tahu Tandang Semarang, Sederhana Kendal, dan Gagak Sipat Boyolali)," *Univ. Diponegoro*, 2007.

[12] R. Slamet, *Pencemaran Air Seri Lingkungan*. Surabaya: Karya Anda, 1984.

[13] W. S. Nursandi S. Mumu; Hartati, Etih, "Perbaikan Kualitas Air Limbah Industri Farmasi Menggunakan Koagulan Biji Kelor (Moringa Oleifera Lam) Dan Pac (Poly Alumunium Chloride)," *J. Teknol. Lingkung. Univ. Trisakti*, vol. 4, no. Vol 4, No 3 (2008): JUNI 2008, pp. 68–73, 2008.

[14] T. T. Pairunan, "Perangkat Lunak Pendukung Keputusan Analisis Pengelolaan Kualitas Dan

Pengendalian Pencemaran Air Sungai," *J. Ilm. Sains*, vol. 12, no. 2, p. 105, 2012, doi: 10.35799/jis.12.2.2012.610.
[15] H. Effendi, *Telaah Kualitas Air : Bagi Pengelolaan Sumber Daya dan Lingkungan Perairan*. Yogyakarta: Kanisius, 2003.
[16] A. T. Sastrawijaya, *Pencemaran Lingkungan*. Jakarta: rineka cipta, 2000.
[17] W. Arya Wardhana, *Dampak Pencemaran Lingkungan*. Yogyakarta: Andi.
[18] A. L. Underwood and J. R. A Day, *Analisis Kimia Kuantitatif (terjemahan A. Hadyana Pudjaatmaka)*, 5th ed. Jakarta: Erlangga, 1999.
[19] P. Sunu, *Melindungi Lingkungan dengan Menerapkan ISO 14001*, Pertama. Jakarta: PT. Gramedia Indonesia, 2001.
[20] C. C. Hach, K. R. L. Jr, and C. R. Gibbs, "Introduction to Biochemical Oxygen Demand," *Nephrology*, vol. 19, no. 7, pp. 1–1, 1997, doi: 10.1111/nep.12265.
[21] M. R. Penn, J. J. Pauer, and J. R. Mihelcic, "Biochemical oxygen demand. Environmental and Ecological Chemistry," vol. II, pp. 278–297, 2009.
[22] Pariwono, "Oksigen Terlarut (DO) Dan Kebutuhan Oksigen Biologi (BOD) Sebagai Salah Satu Indikator Untuk Menentukan Kualitas Perairan," *Oseana*, vol. 30, no. 3, pp. 21–26, 2005.
[23] S. Fardiaz, *Mikrobiologi Pangan 1*. Jakarta: Gramedia, 1992.
[24] R. Achmad, *Kimia Lingkungan*. Yogyakarta: Andi, 2004.
[25] Y. Xia, L. Huang, and L. Xu, "Characteristics of diffuse source N pollution in Lean River catchment," *Procedia Environ. Sci.*, vol. 10, no. PART C, pp. 2437–2443, 2011, doi: 10.1016/j.proenv.2011.09.379.
[26] N. A. D. Putri, "Kebijakan Pemerintah Dalam Pengendalian Pencemaran Air Sungai SIAK," *J. Ilmu Polit. dan Ilmu Pemerintah.*, vol. 1, no. 1, pp. 68–79, 2011.
[27] L. W. Green and M. W. Kreuter, *Helath promotion planning-An educational and environmental approach*, 2nd ed. California: Maytield Publishing, 2000.
[28] R. P. Fisher and R. E. Geiselman, "The Cognitive Interview method of conducting police interviews: Eliciting extensive information and promoting Therapeutic Jurisprudence," *Int. J. Law Psychiatry*, vol. 33, no. 5–6, pp. 321–328, 2010, doi: 10.1016/j.ijlp.2010.09.004.
[29] S. Notoatmodjo, *Metodologi Penelitian Kesehatan*. Jakarta: Rineka cipta, 2005.

Acknowledgments
Author wishing to acknowledge assistance or encouragement from the supervisors and all experts who participated in this study.

AUTHOR INDEX